MORAY HOUSE
AND THE ROAD TO MERGER
An Insider's Account

MORAY HOUSE
AND THE ROAD TO MERGER

An Insider's Account

Gordon Kirk

DUNEDIN ACADEMIC PRESS
EDINBURGH

Published by
Dunedin Academic Press
Hudson House
8 Albany Street
Edinburgh
EH1 3QB

ISBN 1 903765 33 1

British Library Cataloguing in Publication Data
A catalogue record for this book is available
from the British Library

Typeset by Patty Rennie Production, Portsoy
Printed and bound by Cromwell Press

Contents

Foreword

This book completes a trilogy on Moray House. The first, *Moray House and Professional Education*, was an edited collection of papers on aspects of the work of the college. Published in 1985, it was a celebration of 150 years of the institution's involvement in professional education. The second, *Moray House and Change in Higher Education* (1995), was a further collection of papers in which colleagues undertook an analysis of their specialist field and sought to assess institutional progress at a time of significant change in higher education.

If anything, the pace of change has quickened, and now all of Scotland's former colleges of education have been incorporated within universities. Moray House was the first of the colleges of education in Scotland to cross the binary line that separated the universities from other institutions of higher education, by becoming an associated college of Heriot-Watt University in 1991. However, when it merged in 1998 it was with The University of Edinburgh. There has been much speculation about Moray House's change of institutional allegiance. This final part of the trilogy, which charts the institution's progress from independent monotechnic to Faculty of Education of a leading university, may provide some with the explanations they seek.

I am grateful to Douglas Grant of Dunedin Academic Press for all his work in effecting the translation from manuscript to published volume; to Professor Mike Anderson, Dr Bob Glaister and Dr Jim O'Brien, as well as the publisher's anonymous critics, for reading all or part of the manuscript and making many valuable suggestions and comments; to Alice Hamilton, formerly the Principal's Personal Assistant, who safeguarded the files while I worked on them prior to their lodging in the Moray House archive, and who provided all kinds of practical support; and to Mary Speed, who typed and retyped the manuscript and brought to that task that quicksilver dexterity, patience and commitment that I have appreciated now for more than twenty years.

Introduction

One of the features of higher education over the last two decades is the number of institutional amalgamations and mergers that have occurred, principally involving the incorporation of smaller specialist colleges within larger higher education corporations or universities. The formation of such institutional alliances has not passed unrecorded and the literature of higher education now includes reports which offer, with varying degrees of candour, insider accounts of the merger process. Why add to this genre?

The Moray House merger was unique in several ways. When Moray House became an associated college of Heriot-Watt University in 1991, it was the first of Scotland's colleges to breach the binary line that separated Scotland's university system from its centrally-funded institutions of higher education. Besides, each of the two key changes in status – from free-standing monotechnic to associated college of Heriot-Watt University, and then from that to merger with the University of Edinburgh as its Faculty of Education – entailed a significant and unexpected change of direction. The first of these institutional initiatives had a dramatic impact on the educational community in Scotland. The intimation of the press conference at which the institutional linkage was first announced simply stated that Heriot-Watt University would be entering into collaboration with another higher education institution and the media interest was enormous. The disclosure that the partner was to be Moray House was met with complete surprise. Was it not the case that Moray House had a long tradition of collaboration with the University of Edinburgh? Was the Principal of Moray House not the most aggressive champion of the separate existence of the colleges? Had the same Principal not contemptuously dismissed the case for changing institutional status as the "immature hankering after university recognition"? If Moray House had eventually succumbed to the temptation to seek that very recognition, why not with its more obvious partner, the University of Edinburgh, with which it had been in association in one form or another since 1876 and whose Principal for a significant period earlier in the century had combined that role with that of Professor of Education at the university?

The switch from associated college of Heriot-Watt University to merger with the University of Edinburgh as its Faculty of Education represented an even more radical change of direction and was even more unexpected. If Heriot-Watt University was "shocked" at the news that discussions were taking place between one of its associated colleges and a sister university, the rest of the higher education community, not least staff and students at Moray House, were taken almost completely by surprise. Had the Principal of Moray House not persistently and vigorously defended the continued separate existence of Moray House? Did he not as recently as his address to graduates in 1995 defiantly declare that, notwithstanding the prevailing mergermania and notwithstanding the difficult financial climate,

> *we at Moray House are confident in our future financial and academic well-being as an autonomous institution of higher education. We consider that Moray House is an institution with an international reputation; we value and take pride in our institutional independence and we intend to retain it.*

And if it had now been acknowledged that merger represented the most appropriate way forward, why not merge with Heriot-Watt University, the institution with which Moray House had been in close collaboration for six years as one of its associated colleges? And why should merger be sought with the University with which Moray House had pointedly chosen not to collaborate when it sought a closer association with a university in 1989? How are these switches of institutional allegiance to be explained?

Many of these questions have already been posed by commentators on higher education in Scotland. Do the various institutional manoeuvrings provide evidence of a relatively small, and some would say vulnerable, institution being buffeted by forces beyond its control and seeking whatever haven appeared to be available? To what extent did they demonstrate an unscrupulous and calculating opportunism? More improbably, are these moves to be seen as the outworking of a coherent strategy of institutional development and positioning that transferred Moray House from the status of a free-standing monotechnic institution to the Faculty of Education of an internationally acclaimed university?

This book offers answers to such questions. Drawing on documentation already in the public domain and a principal's confidential files, the book analyses the changing alliances against the background of profoundly dislocating changes that beset higher education over the past two decades. Since the book is written from the standpoint of the Principal of Moray House it is certainly an "insider" account. It draws substantially on insights and experiences based on direct participation. At the same time, it also draws on a substantial range of evidence already in the public domain. The adoption of that "outsider" perspective should serve as a corrective to some of the acknowledged dangers to which insider accounts are prone, most notably the temptation to re-write history in the pursuit of self-aggrandisement. To the extent that the outsider perspective is successful in promoting a more detached and measured analysis, the book will realise its principal aim: to offer an essay in explanation rather than to undertake an exercise in self-justification.

1

Independent Monotechnic

Introduction

One of the features of Scotland's educational tradition is that the education of teachers has taken place in separate institutions devoted to that purpose. The establishment of provincial committees for teacher education in 1905 was intended "to ensure that that training shall be brought into as close connection with the university organisations as the attainments of the students upon entering admit of"[1] However, as Stocks[2] has shown, Craik, the Secretary of the Department, had no confidence in the universities as teacher educators. His successor as Secretary, John Struthers, was even more dismissive. It was he who referred to "professors as a body, welled up in their impenetrable fortress of academic seclusion" and ventured the opinion that, if what was needed was a broader conception of education for teachers, "the university is the last agency in the world through which we are likely to obtain it".[3] Accordingly, the Departmental strategy was to insist upon and to secure a differentiation of functions, with universities committed to promoting general education, whilst professional education was the responsibility of specialist training institutions. According to Bell, throughout the century there was "a sustained and deliberate policy on the part of a highly centralised government department, the Scottish Education Department (SED), to keep the control of professional teacher training, as opposed to general teacher education, out of the universities' too independent hands".[4] And Bell's further study of the university departments of education provides a vivid analysis of some of the conflicts which attended that policy.[5]

The separation of teacher education from the universities was reinforced by the division of ministerial responsibilities: the Scottish universities fell within the jurisdiction of a London-based minister at the Department of Education and Science, while colleges of education formed part of the responsibilities of the Secretary of State for Scotland. That arrangement was justified on the grounds that, as the minister responsible for Scotland's schools, the Secretary of State had to control the professional preparation of teachers for these schools.

The Advisory Council Report, *Training of Teachers* (1946)

Outside the Scottish Office, the place of teacher education within the higher education system was a matter for debate in a series of national reviews. The first of these was the report of the Advisory Council on Education in Scotland,

Training of Teachers (1946). Having established a rationale for the education of teachers in an age of reconstruction, the Council directed its attention to the question: Should the training of teachers be taken over by the universities? Several advantages were identified for such a move: it would enhance the status of the profession; it would attract stronger entrants; it would create an all-graduate profession; it would enable all students to have the advantages of participation in the life of the university; and it would be easier to secure the integration of the students' general education with their professional preparation. Magisterially, members of the Council affirmed that they were "not, however, convinced of the soundness of these claims".[6]

In the first place, it was held that the training of teachers "is an integral part of the state educational system", and that it would be undemocratic if it were "entirely controlled by a private body, many of whose members are not in close touch with the schools". Secondly, the distance of the training institutions from the universities would leave them segregated from the life of the university. Thirdly, such a move, entailing as it would "graduation for all", would exalt the intellectual at the expense of other dimensions of education, representing a failure "to overcome our Scottish propensity to over-emphasise school attainment and the intellectual side of education". It would fill our schools with teachers "whose minds had been stiffened in that part of our educational tradition" and would "impoverish" the schools of "many of the non-bookish girls who train as teachers of young children", as well as men and women with experience of the world of industry and commerce.

These counter-arguments were reinforced in the Council's mind by the fact that the proposal to transfer responsibility for teacher training to the universities did not receive significant support: of the seventeen witnesses who gave evidence, sixteen were opposed to the principle of graduation for all primary teachers; and of the thirty-nine who expressed a definite view on the matter, thirty-four were in favour of placing the control of the training of teachers in the hands of ad hoc training authorities rather than with universities. Moreover, the universities themselves were reluctant to accept full responsibility for the training of teachers, since, in the Council's view, it would "endanger the fulfilment of their true cultural mission". The presence of a significant proportion of students with special vocational aims would mean that "the whole character and balance of interests of university life would be materially altered". Finally, if a special degree for teachers were introduced it would be regarded as inferior to established degrees, with the result that there would be a lowering rather than a raising of the status of the profession.

Despite the thoroughness with which the Council addressed the question of the location of teacher education, its opposition to the transfer of responsibility was not unanimous: two of its members, both former presidents of the Educational Institute of Scotland, in a note of reservation expressed their view that "the teaching profession will never attain to that place in public esteem which the importance of its work deserves, nor will it attract in sufficient numbers persons of the high quality which it requires, until it becomes a graduate profession".

However, the Advisory Council's principal reason for rejecting the transfer of responsibility of teacher education to the universities was that it believed that the preparation of teachers for the educational service "is a function of such scope and social importance that it deserves an institution to itself, an institution with a singleness of purpose which no department of a larger institution could ever have". The Council proposed the establishment in each province of an "Institute of Education", which would be a community bound together by "a common central purpose and shared interests", and with a community life deliberately shaped to foster the ideals, attitudes and values of a modern profession. It would be "in the van of educational advance"; "a centre for educational research in all its branches"; "a resort for teachers and the focal point of the educational activities of the province"; and it would be staffed by "men and women of the very highest attainments", comparable in status to university professors, with salaries to match. The Institutes would acquire the right to award degrees and would take over the work of the university departments of education. The Council therefore envisaged the education of teachers taking place in highly specialised institutions that might not have the academic standing of universities but would have a single predominant preoccupation with teaching.

Laudable as the Advisory Council's articulation of the principles of teacher education undoubtedly was, it reinforced the tradition which excluded teacher education from universities. While the Advisory Council's report did not issue in a programme of government action, its significance is that it reflected a professional consensus on the institutional context for the education of the nation's teachers. That task, it was clear, should be the responsibility of separate institutions of inferior standing. The promise of degree-awarding powers implied that the Institutes of Education would need to undergo a significant maturation before they acquired that badge of academic respectability.

The Robbins Report, *Higher Education* (1963)

The legacy of the Advisory Council, that teacher education was a non-university and lower status activity, was bequeathed to subsequent national enquiries. The Robbins Report of 1963[7] responded to the finding that some 40% of students in colleges of education had university entrance qualifications, and to the evidence on the need for an all-graduate profession, by proposing a BEd degree, to be developed "in partnership" between the colleges and the universities. The new degree was intended for a minority of college of education students – those judged to be the ablest – and when the new programmes were introduced, first at Aberdeen and then in the other colleges from the mid-1960s, they had some striking features. Apart from the fact that they were restricted only to a minority, they tended to be modelled on the established Scottish Ordinary degree, with seven graduating courses, mainly of the traditional academic kind, with professional studies and teaching being almost completely marginalised. Indeed, when in due course the students completed their studies they acquired the BEd degree at the university graduation ceremony and attended the college to receive their teaching qualification in a separate ceremony. That arrangement certainly was in accord with the Robbins egalitarian commitment to extending degree opportunities, but it did little to enhance the standing of teacher education as a university-level study. Indeed, if anything, the Robbins Report had the opposite effect: it reaffirmed the academic inferiority of the colleges by recommending, mainly in response to such advocates of the universities as Walker of Aberdeen and Adams of St Andrews, that graduates on the one-year teacher education programme could escape "the less stimulating academic atmosphere" of the college by following courses in education and psychology at the university. That recommendation, as well as the recommendation on the BEd, served to confirm the separate standing of the colleges as principally training establishments operating at an academic level below that of the universities.

The House of Commons Select Committee (1971)

A further test of professional opinion was provided by the work of the Scottish Committee of the House of Commons Select Committee on Education, which visited Scotland in the course of 1970 and took evidence widely throughout the educational community. Invariably, many of the exchanges concerned the BEd. While some of the evidence indicated that the degree had supporters, particularly staff who enjoyed the challenge of working at the university level, several substantial reservations were expressed. It was clear that some of the colleges found the relationship with a neighbouring university too constraining. For example, Wood of Jordanhill would have preferred "courses designed for the job we have to do . . . and not designed in a pattern that was conceived perhaps two hundred years ago". Staff regretted the divisiveness that had been introduced by the Robbins BEd, which separated degree students from others and differentiated between those staff who were "recognised" teachers of the university and those who were not. Finally, while the degree was intended to promote institutional development, it tended to undervalue professional studies and, if anything, induced an attitude of dependence on the universities.

Of course, the answer to the question of divisiveness was to introduce a degree for all students and that lay at the heart of the evidence submitted by Wood, who expressed himself "not opposed to university control of teacher training at all, provided it will take in the whole of teacher training and not just the best students or the ones that the university would like".[8] However, even if that principle had been conceded, there was strong disagreement on the nature of that degree. For some, it would have to be a professional degree in which teaching and other professional studies would receive clear recognition. However, for many teachers a special teachers' degree would not enhance the standing of the profession. It would be regarded to be of lower quality than the universities' traditional offerings. That lay behind Woods' discouraging comment that "the staff are teachers whose own profession will not support them", a reference to the opposition of the teachers' professional associations to degrees other than those of the traditional academic type. Wood repeated that he supported the idea of degrees for teachers, even if these were independent of the university, but added, "That is my own personal view, but I am not a Scot."[9]

That characteristically sardonic remark brilliantly encapsulated the tension at the heart of teacher education in Scotland: it acknowledged that teaching imposed demands that were sufficiently challenging to merit the award of a university degree; that a degree for teachers had to reflect professional concerns in a way that the traditional university degree did not; and that a degree that differed too radically from the traditional pattern, far from earning the respect of the teaching profession in Scotland, would be regarded as inferior. As long as that disagreement persisted, as long as it continued to be disputed that teacher education fully merited the award of a university degree, there was justification for the maintenance of a separate college of education system.

Institutionalisation of the Monotechnic Principle

There were two factors, then, which reinforced the monotechnic principle. Firstly, the existence of a separate teacher education system was politically necessary, enabling the Secretary of State for Scotland to exercise that degree of control that was assumed to be necessary to ensure that teachers were trained in line with the changing requirements of the schools. The incorporation of colleges within universities, which were answerable to a separate minister, and a London-based one at that, was thought to endanger the Secretary of State's responsibility for the Scottish education system as a whole. Secondly, there was a professional consensus, extending over many years, that teacher education was not an appropriate university study, either because teaching required professional capacities that would not be reflected in a traditional university degree, or because the level of academic study required to be a teacher was judged to be lower than that required for a degree. These arguments provided a political as well as an educational justification for differentiating teacher education from the rest of higher education, and that differentiation was institutionalised, not just in separate colleges of education but in a separate college of education sector with its own forms of financial control, staffing and collective bargaining, and academic accountability, all administered and regulated with all the apparatus of political control by a separate branch of the Scottish Office.

The Council for National Academic Awards

The Council for National Academic Awards (CNAA) played a crucial role in strengthening the monotechnic tradition. Established in 1964 to award degrees to students who completed approved programmes in non-university institutions, the CNAA evolved a distinctive and rigorous approach to validation which had a transformational impact on institutions. Its basic philosophy had three key features. First, it was accepted that the quality of a programme depended on careful scrutiny by a knowledgeable and experienced group of critical outsiders drawn from all parts of the UK. Secondly, free from traditional conceptions of a university degree, it encouraged institutions to evolve teacher education programmes based on a functional analysis of teaching itself, fully recognising the centrality of school experience and committed to securing the full involvement of teachers in the planning and delivery of courses. Thirdly, acknowledging that strong courses cannot flourish unless they are located in an appropriately supportive environment, the CNAA expected institutions to undergo, through its institutional review procedures, a full scrutiny of their quality assurance and quality control systems. There is no doubt that the existence of the CNAA eliminated the dependence of colleges on universities, enabling them to obtain degrees for their students without compromising their professional standards and obligations, and without having to subscribe to degree structures that took insufficient account of professional requirements. The CNAA was therefore in many ways the protector of the academic and institutional independence of the colleges.

An All-Graduate Profession

In February 1983 the Secretary of State announced the introduction of a new BEd degree for primary teachers, a measure which created an all-graduate profession in Scotland. Tempting as it may be to attribute this government initiative to the strength and persistence of the representations made by colleges, the General Teaching Council, and others, the change was virtually inevitable once a degree for primary teachers was introduced in England. At the same time, the Secretary of State intimated that a working party would be established to develop guidelines for the new degree. The guidelines that emerged ensured that the award would differ radically from the BEd introduced after the Robbins Report, a degree which, modelled on the traditional university pattern, marginalised professional studies and school experience. By contrast, the new degree was required to be professional in orientation, to assign a central role to school experience and professional studies and, in general, to reflect those features of professional programmes which were strongly supported by the CNAA. While some members of the working party were anxious to see the new degree provided only by universities, the case for CNAA validation was also strongly argued and, ultimately, colleges were allowed freedom to choose their own validating body. Had CNAA validation not been made a possibility, it is highly unlikely that the national guidelines would have had anything like the same strong professional emphasis.

These developments – the enhanced role of CNAA in course validation and the introduction of the new Primary BEd – contributed significantly to the institutional development of the colleges. The extension of degree opportunities in music and in technology meant that by the mid-80s the whole undergraduate population of the colleges was engaged in degree-level programmes. At the postgraduate stage institutions began to extend the range of in-service awards for practising professionals, many of these at master's degree level. In addition, in response to the challenge of operating at degree and postgraduate levels, institutions came under pressure to ensure that their programmes were adequately underpinned by research and development work. Fortunately, there was a significant extension of these activities. Supported by a specific staffing allowance for research, colleges were able to enhance their research profile. In addition, many members of college staff were heavily involved in national curriculum development projects and in supporting these at local level. Indeed, across the system as a whole, some 200 full-time equivalent posts were protected for in-service activities, thus ensuring that the close links the colleges had developed with the teaching profession were maintained and strengthened.

Contraction of the System

These significant academic developments in the colleges took place against a background of retrenchment and financial stringency, partly necessitated by a decline in the school population. A direct consequence of that decline was that in 1979–80 there were significant prospects of teacher unemployment, with the GTC Supply Committee estimating that only 78 primary teachers could be employed from an output of 703, and 991 secondary teachers from an output of 1,684.[10] In response to the over-production of teachers, the SED proposed intake figures for 1980/81 which meant a decrease of 29% in the primary diploma, 35% in the primary postgraduate course, and 16% in the secondary course over 1979/80. These figures would have led to a total number of students in all courses in all colleges of 5,967 against an estimated college capacity of 14,450.[11] Obviously, that was an embarrassing prospect for a Conservative government for it would place ministers in a position where they had to contemplate the closure of colleges when, only three years before, they had vigorously attacked the Labour government for making proposals precisely of that kind.

However, the GTC, supported by a recommendation from its Supply Committee, came to the rescue. That committee had recommended that "there should be a reduction in the number of colleges engaged in training teachers". That recommendation was clearly motivated by a desire to retain the larger colleges because of their "ability to diversify, to support specialist facilities, to foster educational research and development, and to provide centres of excellence for the educational system as a whole"[12], and the GTC duly communicated that view to the Scottish Office. The government responded to that recommendation in August 1980 by announcing the closure of two colleges – Callendar Park College of Education and Hamilton College of Education – and the merger of two others – Notre Dame College and Craiglockhart College. Despite a vigorous campaign by all of the colleges concerned, and in the face of the most intense political pressure, on this occasion the government stood firm and the 1981/82 session opened to a seven-college system. However, the student number projections continued to offer grounds for concern. The SED projections for primary represented a further reduction in overall student numbers from 2320 to 1600 for 1983/84, and the corresponding figures for secondary were from 3110 to 1760. Indeed, the longer term projection for secondary was 450.[13]

One illustration of the extent of the decline concerns physical education. When the two degrees in physical education were established in 1975, the intake to the Scottish School of Physical Education at Jordanhill was 100 students per year and, because of the apparently higher loss of female physical education teachers from the profession, 165 at Dunfermline College of Physical Education College, a total of 265. In 1982/83 the combined intake to physical education programmes for the whole country was 40.[14]

The vulnerability of the colleges was emphasised by several other embarrassing factors. Firstly, the decline in staffing did not keep pace with the decline in student numbers. Thus, in 1975/76, there were just over 15,000 students, and that number had reduced to 7,072 by 1983/84, a drop of 53%.[15] On the other hand, the number of staff dropped only from 1432 to 839 over the same period, a drop of 41%.[16] Secondly, the staff:student ratio of 1:8.5 in 1983 compared very unfavourably with the figure of 1:11, which was the figure used by the Department of Education and Science (DES) for comparable institutions in England. The application of the DES ratio to Scotland in 1983/84 would have required a further reduction of 132 staff. Thirdly, it was well known that many of the student groups in the colleges were extremely small, particularly in those colleges offering a BEd degree in association with a neighbouring university. Students on that programme had the same rights of choice as other university students and colleges were under an obligation to ensure that student choice was met. The result was a relatively small number of students studying, in groups of two, three, or four, an extensive range of options. In all, there were 101 groups. Of these, 50 consisted of fewer than ten students.[17] Fourthly, SED had evidence of a substantial pool of unemployed teachers. While the exact figure was never known, repeated reference was made to it in the annual consultation paper on intake to the colleges. Indeed, when the Public Accounts Committee (PAC) came to interview James Scott, Secretary of SED, in June 1985, he was put under pressure to explain why it was that somewhere between 30% and 40% of the output were still without posts two years after graduation. Finally, the colleges of education fared badly when their unit costs were compared with those of the central institutions. In 1981/82, the unit cost in a college of education was £3,800, whereas in the central institutions it was £3,062.[18]

The central difficulty facing the colleges was that they were embarked on a major phase of development, not least in the provision of in-service support for schools, but against a background of intensifying financial pressure, the progressive reduction of staff, and the virtual cessation of staff recruitment. While staff responded to the new challenge energetically, morale was not high. The seven remaining institutions, aware of the growing discrepancy between their total student numbers and their overall capacity, and aware also of the political unacceptability of diversification, expected still further closures. The vulnerability of the colleges was made even more acute by the fact that they were small, specialised institutions. It was hardly surprising that a director of education could publicly refer to the college of education system as "languishing"[19] and that another commentator referred to the colleges as "beleaguered".[20] A careful analysis of their function and of their relationship to other institutions appeared to many to be long overdue. That analysis was to be provided by the Scottish Tertiary Education Advisory Council review (STEAC).

The STEAC Review

When it was first mooted in July of 1983, the STEAC was intended to advise the Secretary of State "on the development of non-university tertiary education". It was clear that that review was initiated at least partly because of the parlous state of the college of education sector as a whole. However, when the Principal of Aberdeen University intimated an interest in the merger of all three higher education institutions in Aberdeen, the Secretary of State had a justification for extending the remit of STEAC to include "the general principles which should govern relationships between universities and other institutions". The review therefore created the opportunity for another fundamental appraisal of the institutional context of teacher education.

The evidence from the college of education sector, supported by teachers' unions and others, firmly favoured the retention of separate colleges of education, although three of the college of education submissions suggested that these colleges should either have an association with a university or form part of a regional federation of institutions. The oral evidence from the college principals confirmed their endorsement of the monotechnic principle, and the supplementary evidence submitted by the two principals unable to attend the meeting with STEAC on 29 July 1985 – Carroll and Kirk – was even more vehement in advocating the institutional identity and integrity of colleges of education. The key arguments used to justify that position were that specialist institutions were a source of educational and professional strength; that the strong links the teacher education institutions had developed with the teaching profession needed to be maintained; and that absorption in a larger institutional framework would marginalise teacher education and threaten standards. It is clear that the Council was very favourably impressed by the evidence it received. It acknowledged that the teaching profession believed that the system had "served Scotland well in the past and will continue to do so".[21] It considered that "the peculiar strengths of the colleges of education derive very largely from their single-minded commitment to the teaching profession, which provides a clear focus for their activities, strengthens their sense of purpose, and provides a source of motivation for students".[22]

Apart from a paper prepared by the SED, which sought to focus the attention of the committee on a number of key issues concerning the college of education system, two papers provided the basis for the Council's work on the college of education system: one by Bone, Principal of Jordanhill, and one by Cuming, Principal of Dundee Institute of Technology. Bone's paper acknowledged that the monotechnic character of the colleges was now "unusual throughout the world"[23] and that, if the STEAC was proposing changes in the rest of higher education, it would be "rather strange to leave the colleges simply as they are". He identified a number of options, indicating that his own preference was for full absorption of the colleges of education into the universities "for the next generation of staff and students because of the established reputation of the universities, the opportunities for teaching and research at high levels, and the better conditions of service". His second choice would have been associated status with the universities, "if it was regarded definitely as a transitional stage to full absorption", but, "as a permanent arrangement, it would attach inferior status to teacher education in a way that I would regard as unacceptable".[24] He was aware that his own preference would not be universally acceptable: "There would be objections . . . partly because the colleges would lose their identities and partly because, under present circumstances, universities might be unwilling to absorb over-staffed colleges".[25]

While Cuming's paper was concerned with the whole non-university sector, the case he was defending – the need to establish a strong non-university sector to rival the universities – necessitated reference to the colleges of education. His comments were not flattering, referring indeed to "the infirmities of the colleges of education which arise from their academic isolation, over-provision, and excessive costliness".[26]

The case for reform certainly had its supporters on the Council. Williams of the University of Glasgow was totally unimpressed by the arguments for the status quo. He "accepted that the Secretary of State would wish to retain control over the output of teachers, but did not regard this as a reason for excluding the possibility of teacher training being done by universities, where the government exercised effective control over intakes of medical and veterinary students". He saw no "insurmountable difficulties" in universities providing teacher education "even for Roman Catholic students",[27] an interesting observation in the light of the subsequent merger of St Andrew's College of Education with the University of Glasgow in 1999. In addition, Cuming, whose critique of the colleges of education system had been noted, supported the case for change, maintaining that "in small institutions the staff were academically isolated, like the students" and the "costs of administration and infrastructure were also unduly high".[28] He also referred to "substantial unused accommodation which could be used by other institutions".[29] In a letter to the Chairman he made reference to the claim that teacher education is best undertaken in monotechnic institutions and maintained that he had "yet to see any evidence for the validity of that proposition".[30] And the Chairman of the Council himself implied that he, too, favoured reform when he "expressed disappointment at the reluctance of the college of education principals to accept a reduction in the number of monotechnic colleges or the possibility of union with central institutions",[31] and when he claimed that "teachers were probably too isolated from the world outside education".[32] With such a formidable group in favour of change, why did the Council conclude that "teachers should continue to be trained in specialist institutions"?[33]

Firstly, there was fear of academic drift: members of the Council opposed absorption of the colleges into universities "because of the likely adverse effect on professional training through an excessively academic approach".[34] One member of the Committee made the same point rather differently, but also controversially, when he stated that the involvement of colleges with universities "might be of advantage to student teachers as students but it would not necessarily help them to become better teachers".[35]

Secondly, despite Williams's assurances that the Secretary of State's right to control numbers would be protected in the same way as for medical students, and despite Bone's reminder that teacher education was incorporated within the universities in England and Wales, members of the Council clearly feared that the constitutional position in Scotland, where teacher education was the responsibility of a different minister, would weaken the control which ministers at the SED had always exercised over teacher education.

Thirdly, members of the Council feared that if, as Bone claimed, the incorporation of colleges within universities would lead to the adoption by college staff of university pay and conditions,[36] the costs of training teachers, already high because of the favourable staff:student ratio, would increase still further, and that was confirmed by Williams's claim that "universities would not be able to take on all staff in colleges of education and would only accept the number which could be employed at university rates for the sum of money that was transferred from the colleges".[37]

Fourthly, it was acknowledged that incorporation of colleges of education in another institution would not necessarily mean integration, in the sense of enabling potential teachers to rub shoulders with those destined for other occupations. For one thing, the central institutions were already full to capacity; for another, "where a college of education became part of another institution of higher education but remained on its present site, students and staff would continue to be segregated".[38]

Fifthly, the Council was clearly impressed by the unanimity of view expressed by the principals in their oral and supplementary personal evidence, that display of unity being enhanced by the absence, because he was a member of the Council, of the one principal most committed to linkage with the universities.

Finally, the group on the Council committed to change was divided, principally because Cuming was angling for incorporation of the colleges not with the universities but with the central institutions. The paper Cuming submitted argued for a strong public sector organised on a federal basis to create a new University of Scotland. It would have been contradictory for Cuming therefore to support the incorporation of the colleges of education within the universities. Besides, he was explicitly opposed to that solution for two additional reasons: firstly, since the public sector "specialised in the provision of vocational courses", he did not believe "it would be appropriate for teacher training to be transferred to the universities"; secondly, he maintained, "some departments of education in Scottish universities were very weak and the absorption of colleges of education into university faculties of education would not solve the problem of academic isolation".[39] Finally, drawing on every conceivable argument to keep colleges away from the universities, he pointed out that "absorption of teacher training by universities would end validation of such courses by the CNAA and would be unacceptable to some colleges of education".[40]

All of these considerations, together with the evidence that strongly supported the college of education system, persuaded the Council that

> *The standard of teacher training in Scotland will, in our view, be best preserved by its concentration in thriving specialist establishments with a common sense of purpose.*[41]

Having decided on the institutional context of teacher education, the Council moved to address the question of the size of the college system. The Council had been made aware that there was substantial over-capacity and there was considerable pressure on the Council for action. At a time when resources were scarce, when it was vitally important to assure that these were deployed to best advantage, the notion of over-staffed and over-resourced colleges of education was bound to be an embarrassment to the SED.

One of the complicating factors was that, despite the acknowledged need for rationalisation, there was a need for dispersal of provision to ensure that teachers had access to colleges for in-service purposes. The Department's answer to that need was to invoke the principle of out-stations, which had enabled St Andrew's College to provide a nation-wide form of support for Roman Catholic teachers in Edinburgh and Dundee. Given the pressure under which the Scottish Office had been placed by the Public Accounts Committee report, STEAC was bound to favour the rationalisation of the system, while seeking to maintain a degree of dispersal. It recommended as follows:

> *To achieve adequate geographical coverage in the training of teachers for non-denominational schools, colleges of education should be retained in the north, west and east of the country. A total of four colleges might, therefore, be a more appropriate number to cover the pre-service training needs of the Scottish teaching force.*[42]

That rough calculation was to provide the basis for the most sustained debate once the report was published. Indeed, the consultation on the STEAC report was dominated by the college of education issue and, in an attempt to reconcile all of the conflicting pressures, the Secretary of State adopted the "no closure" option, whereby the total number of colleges would be reduced, while no centre would be closed. The Secretary of State therefore appeared to endorse the central finding of the STEAC report about the need for a separate college of education sector. There was room for some satisfaction in the colleges. Despite the enormous difficulties they had been living through, the value of their work had been publicly endorsed and their place as a separate sector in higher education had been clearly confirmed.

Moray House and the Monotechnic Principle

Moray House played a significant part in the development of the thinking about the future of the college of education system and the defence of the monotechnic principle. Three contributions, in particular, may be highlighted. Firstly, Moray House became known as the staunchest supporter of CNAA validation. Its principal had been an active member of the CNAA for five years by the time of his appointment, shortly after which the college embarked on a major programme of course development with CNAA as the sole validating body. As more staff became involved in CNAA work, and in response to two institutional reviews, Moray House radically altered its whole mode of operation, establishing rigorous and incisive quality assurance arrangements. In the debate which *The Times Educational Supplement Scotland* opened following the establishment of the Scottish Council for the Validation of Courses for Teachers (SCOVACT), an attempt by the universities and colleges to collaborate in a Scottish-based system of course approval, Moray House's principal was portrayed as the champion of CNAA validation, advocating its clear superiority over SCOVACT as a validating body and taking issue with the critics who sought to undermine the CNAA in favour of its somewhat homespun and inexperienced rival.

CNAA validation was thought to confer several advantages. In the first place, it was based on peer group evaluation, "the rigorous and independent scrutiny by a group of knowledgeable and experienced outsiders". It enabled an institution "to key into a national network of expertise" and, in that way, to access best contemporary practice.[43]

Secondly, in comparison with regional validation arrangements such as SCOVACT, in which "close personal and institutional ties . . . stand in the way of free and frank exchange", CNAA validation was objective "in the sense that it is undertaken by people who are outsiders . . . (who) have no axe to grind, and who, having no connection with a submitting institution, can therefore speak with impunity on a course proposal and judge it fairly on its merits".

Thirdly, again contrary to SCOVACT practice, CNAA validation entailed the scrutiny of the institutional context in which a course was located. Based on the assumption that strong courses cannot flourish in an inappropriate academic environment, the CNAA institutional review procedures involved a "visit to the institution by a group of CNAA members, again drawn from a variety of institutions, who are skilled in institutional analysis and who have to be persuaded by the evidence provided that the college is able to sustain the course and foster its development". The combined effect of course and institutional review was acknowledged "to contribute a powerful stimulus to institutional and staff development".

Finally, it was maintained even in university circles that "course validation of this kind is altogether more thorough and systematic than is usual in universities".

It was necessary also to defend CNAA against its detractors. One of the most frequent lines of attack was that "as a London-based agency CNAA is ill-equipped to validate courses in Scotland", and that it was necessary to avoid "the penetration of Scottish education by an English body". The article observed that, ironically, such comments tended to be made by "ex-public school, ex-Oxbridge people currently on the staff of our ancient Scottish universities". In an effort to defend CNAA against the charge of being "an Anglicising quango", it was maintained, firstly, that "while there is an obligation on CNAA to familiarise itself thoroughly with education in Scotland, there is a corresponding obligation on institutions to clarify the professional circumstances into which their students will be moving, and also to demonstrate precisely how the proposed course would equip students with the skills and understandings for their professional work". Secondly, it was affirmed that "the principles on which sound programmes of teacher education are based are not Scottish or English: they are professional". One of the most serious threats to high quality courses, it was claimed, "is the parochialism that insists that we have little, if anything, to learn from others in educational matters. That is all the more reason for a form of validation that is external." The final sentence was a further jibe at the parochialism that might occur when course validation and review were undertaken by SCOVACT, or where, indeed, external validation appeared to be unknown, as was the case in most universities.

Advocacy of that kind had a number of effects: in extolling the advantages of CNAA it defended the monotechnic principle; it implied that colleges in association with CNAA were accustomed to quality assurance arrangements that were in advance of those then operating in universities; and it deepened the division between colleges of education and the universities.

Secondly, Moray House featured prominently in the working party on the new BEd degree and in the discussions that heralded the introduction of the new award. Within a few weeks of the Secretary of State's announcement, *The Times Educational Supplement Scotland* carried an article under the title *The Unashamed Professional*, drawing on the opening sentence, which claimed that the purpose of the article was "to urge that the new degree should be fully, explicitly and unashamedly professional: it should be preoccupied with the cultivation of competent professional performance".[44] The central thrust of the argument was to maintain that, in professional education, theoretical understanding was not a form of cerebral virtuosity but, rather, was "the kind of understanding that issues in intelligent action and sustains reflective practice". That professional focus proved too much for one critic, who maintained that such an approach was narrowly conceived and demonstrated that the writer had professional blinkers. He went on to regret that a professional degree would impel colleges of education "to look inwards and downwards to primary schools for their inspiration instead of outwards and upwards to other places of learning for academic stimulation".[45] The rejoinder to that criticism could not have been more blunt:

> *That judgement is symptomatic of many of our current shortcomings in teacher education in Scotland. It reflects an immature hankering after university recognition; it fails to acknowledge that professional problems impose demands that are as absorbing, as intellectually challenging, and as educative as anything offered by conventional undergraduate programmes; it demeans the professional activities and expertise of college of education staff; it seeks to divert the attention of such staff, by a straight appeal to academic snobbery, away from their primary and continuing focus of concern – the schools themselves; and, finally, it almost completely discounts the professional expertise of classroom practitioners as a legitimate knowledge base for those aspiring to be teachers.*[46]

The article went on to congratulate the Secretary of State on the quality of the guidelines, maintaining that these offered "the opportunity for a fresh start and for a reaffirmation of the importance of the professional over the merely academic".

That stipulative distinction between the professional and academic lay at the heart of subsequent discussion. It was acknowledged that a degree for teachers, like any other degree, should be academically demanding: it should seek to cultivate the general powers of the mind. The argument that was being advanced was that, as well as meeting the academic criterion, the new programme should issue in competent professional performance. That is, it was not enough to be the possessor of knowledge: it was necessary to be able to exploit that knowledge to have an impact on people's learning. It was maintained, besides, that such work was educative and did not consist of a narrow vocational training. Critics, notably Humes, questioned that distinction and in a sustained critique[47] maintained that the whole rationale was built up by assertion rather than by logical analysis. Whether or not the Humes strictures were justified, there is no doubt that the new BEd that developed in Scotland was strongly influenced by that conception of a professional degree and was very much in evidence in the accreditation procedures adopted by the General Teaching Council.

Thirdly, Moray House played a leading part in defending the monotechnic system in the context of the developing "mergermania" in higher education in Scotland. That received a significant impetus from the publication in October 1988 of a proposal by Dr Bill Turmeau, Principal of Napier, for the restructuring of higher education. Turmeau's proposals were vigorously attacked in an article in *The Times Educational Supplement Scotland*.[48] Referring, dismissively, to the T-plan, it argued that to fix the optimum size for institutions at 5,000 to 10,000 students was extremely questionable. By the standards operating in the United States, for example, an institution of that size would be extremely small. If there was substance in the argument about economies of scale, presumably it would apply to institutions of the size being advocated by Turmeau. Moreover, the T-plan was restricted to the non-university sector. Turmeau appeared to be advocating the establishment of a binary system in Scotland at a time when the universities were becoming more and more responsive to national needs. The binary system itself was suspect, for it assumed that there was an essential distinction between universities and the rest of higher education. On the contrary, it was argued that the establishment of a binary system would create serious inefficiencies. It was concluded that

> *For these reasons, the T-plan does not provide a rational basis for change in higher education in Scotland. It is structurally incoherent; it is inconsistent and incomplete in its analysis of the optimum size of institutions; and, by uncritically accepting an out-moded binary system, it stands in the way of the kind of rationalisation it is itself advocating.*

Fourthly, Moray House's impact on the STEAC report was significant. The evidence from its Board of Studies, like the subsequent response to the report's recommendations, was a sustained piece of advocacy on behalf of the college of education system in general, and of Moray House in particular. It envisioned colleges of education as integral features of large federations of institutions set in different parts of the country but with specialist provision maintained for teacher education and other professional fields. In its response to the STEAC report, and in particular to the recommendation that the number of colleges should be reduced from seven to four, Moray House – like other colleges – engaged in a blatant piece of special pleading on behalf of the institution. One commentator had maintained that the STEAC recommendation of a reduction of the system from seven colleges to four would lead to bitter rivalry "inviting colleges to enter an exercise of self-justification and survival of the fittest in order to stave off the axe".[49] Moray House's response was to argue strongly for the monotechnic principle and for the retention of all colleges, mainly on grounds of the dispersal principle. It nevertheless thought it prudent to conclude its submission with a justification for the future of Moray House. To guard against Moray House's case being allowed to go by default, there followed a list of the features of the institution which were thought to commend it: its location in a capital city; its commitment to external validation; its partnership with schools; its standing as an established centre of research and development; its status as one of the largest centres for overseas education in the UK; its programmes in Community Education and Social Work; and its record over the past ten years as having the lowest unit cost of the Scottish colleges. The submission hoped that these qualities justified for Moray House "a secure place in the future of professional education in Scotland".[50]

However, the most vigorous defence of the monotechnic principle was included in the personal submission made by the Principal in response to the invitation from the Council to those who had been unable to attend the STEAC meeting with the other college principals. That personal submission took two forms. The first was a copy of a chapter from *Moray House and Professional Education*, a collection of essays published to celebrate the college's 150th anniversary. The last chapter of that book speculated on the future context of professional education and advanced a number of possible lines of development. It also contained a strong justification of the monotechnic principle:

> *There is ample room in that variety of provision for institutions which make no claim to comprehensiveness of academic or intellectual coverage but which have a distinctive and specialised professional function to perform. A college such as Moray House would be an institution of this kind. Its whole tradition has shaped it as an agency for teaching, research and development in professional education. The centrality of that preoccupation with professional activity does not weaken the institution: on the contrary, it is the source of the institution's strength, providing a focus for individual, interdepartmental and institutional activity. Besides, the relatively restricted range of its academic and professional coverage helps to maintain the quality of the academic environment, to sharpen the sense of institutional purpose and to provide a powerful source of student motivation.*[51]

Turning to the question of institutional amalgamations, it was conceded that teacher education could be performed in universities and polytechnics. It was also acknowledged that there were many colleagues who would happily exchange their existing conditions of service and status for those of a member of staff of a university. However, there were two major difficulties with institutional amalgamations of the kind commonly advocated. In the first place, it was maintained that when the college of education function is performed within a larger institutional structure "it is frequently considered to be peripheral and to fare badly in the competition for funds and resources with other elements". Secondly, it was maintained that the amalgamation of the teacher education function within larger institutions

> *would entail the transference of decision-making on courses and related matters from those who have the necessary expertise and experience to make relevant professional judgements to those who lack that experience and who, in my experience, are inclined to regard education – as opposed to the pursuit of scholarship – with something approaching disdain and who have not grasped the distinction between academic accomplishment and the capacity for intelligent professional action. If the college of education function were to be absorbed within a larger institution professional decision-making would be even further distanced from the professions with whom colleges of education now work very closely and the full professionalisation of teaching and related activities would be inhibited.*[52]

Through advocacy of that kind, Moray House demonstrated that it was not simply a leading monotechnic institution but also one that sought to play a vigorous role in defending the continuation of a separate teacher education system.

The Passing of the Monotechnic

Despite the endorsement of a separate college of education system provided by STEAC, within a decade or so the monotechnic tradition has passed away, with teacher education now integrated into universities. How is that radical transformation in the structure of higher education to be explained? Several factors can be adduced.

Firstly, the STEAC settlement was an extremely fragile one and contained within it the seeds of subsequent institutional unrest and dislocation. Members of the STEAC were well aware that the monotechnic was "unusual throughout the world"[53] and there were those, such as Cuming and Williams, who disputed the claim that teacher education was best conducted in monotechnic institutions. That argument was supported by many bodies during the consultation period – for example, the CNAA, the Convention of Scottish Local Authorities (COSLA), the Association of Directors of Education in Scotland (ADES), and the university departments of education. Besides, some of the arguments advanced in the course of STEAC in favour of the monotechnic principle were not particularly robust. Members of the Council must have been aware that in other parts of the world teacher education was already part of the life of universities. It was also unconvincing to claim that the incorporation of teacher education within universities would lead to academic drift, since the Secretary of State, using his authority to approve teacher education courses, would be able to prevent that. Furthermore, the Secretary of State, in making his announcement to the House of Commons in July 1986, seemed to be not wholly committed to the monotechnic principle, intimating that he was "not ruling out entirely the possibility of some other arrangement if circumstances appeared to warrant it".[54] It appeared as though the STEAC had allowed itself to be over-impressed by the rhetoric of the protagonists of the monotechnic principle.

Given these uncertainties, it was inevitable that the years following the STEAC would be characterised by continuing instability. Student numbers continued to decline; staffing numbers were progressively eroded; and the issue of institutional roles reappeared. It emerged in the first place in the review of the University of Stirling's Department of Education by the Scottish Office. That review concluded:

> *The evidence of our review showed that the Department of Education at the University of Stirling produces well-trained, enthusiastic probationer teachers; provides highly regarded, innovative in-service training courses; and has undertaken valuable classroom-based research.*[55]

That conclusion, which contradicted the STEAC conclusion, was bound to raise questions about the future of the monotechnics.

The colleges were further destabilised by a number of initiatives in which merger was mentioned. For example, in his address to principals and governors in 1989, the Minister dwelt on institutional collaboration, "perhaps even to the point of mergers", and made specific reference to the colleges of education. Secondly, the principal of Napier Polytechnic produced a paper in which he quite specifically argued for the reduction in the total number of Scotland's twenty-five institutions of higher education to ten.[56] The paper proposed an amalgamation of Napier with four other colleges in the Edinburgh area, including Moray House.

A further dislocating factor concerned the future of the CNAA. That fell to be reviewed following the incorporation of the major polytechnics. It was widely feared that, since they had accounted for 85% of CNAA validation work, CNAA would be under threat and that the smaller HEIs would have their work validated by polytechnics. Of course, Moray House could have turned to SCOVACT, the consortium of colleges and universities established in 1983. However, that body restricted its work to the validation of courses for teachers and did not involve a scrutiny of the quality assurance procedures of the institutions. It was therefore not as attractive a body as CNAA had been.

A combination of factors, then, contributed to the decline of the monotechnic principle: the STEAC settlement was an extremely fragile one and was far from establishing a stable environment for colleges of education; the instability was accentuated by the new regime in which institutions were under pressure to secure their own future viability and well-being; there appeared to be a determination to recreate in Scotland a binary system which could be seen as frustrating the institutional aspirations of some colleges by locking them into a clearly identified non-university sector; the leader of the geographically closest polytechnic-type institution to Moray House appeared to be lobbying for large-scale institutional amalgamations; and, finally, the body which had been a main protector of the independence of the colleges of education – the CNAA – appeared destined for closure. That combination of circumstances was bound to create a degree of insecurity for colleges of education, which could be seen as extremely vulnerable institutions. As the first of the colleges to break ranks, Moray House judged that the most appropriate response to the prevailing uncertainty was to establish a linkage which would ensure that its students graduated with university awards. At the same time, it was believed that other educational and institutional advantages might accrue through being attached, albeit loosely, to a larger and more settled and secure institution.

References

1. Minute of the Committee of Council on Education in Scotland, 30 January 1905.
2. John Stocks, "Broken Links in Scottish Teacher Training" in *Scottish Educational Review*, Volume 18, No 2, November 1986, pp 110–120.
3. Marjorie Cruickshank, *History of the Training of Teachers in Scotland*, University of London Press, London, 1970, page 145.
4. R E Bell, *Educational Studies in the Scottish Universities, 1870–1970*, PhD Thesis, Open University, 1986, Vol 1, page 3.
5. R E Bell, "The Education Departments in the Scottish Universities", Chapter 8 of *Scottish Culture and Scottish Education*, edited by Hamish Paterson and Walter M Humes, John Donald, Edinburgh, 1983.
6. Report of the Advisory Council on Education in Scotland, *Training of Teachers*, 1946.
7. The Robbins Report, *Higher Education*, HMSO, London, 1963.
8. Select Committee on Education and Science (Scottish Committee), Session 1969/70, Teacher Training, Minutes of Evidence, London, HMSO, 16 March 1970.
9. Select Committee on Education and Science (Scottish Committee), Session 1969/70, Teacher Training, Minutes of Evidence, London, HMSO, Evidence 16 March 1970.
10. Willis Marker, *The Spider's Web? Policy-making in teacher education in Scotland, 1959–1981*, Sales and Publications, University of Strathclyde, 1994, page 118.
11. Ibid, page 118.
12. Letter from General Teaching Council for Scotland to Secretary of State dated 7 March 1980.
13. Willis Marker, *The Spider's Web? Policy-making in teacher education in Scotland, 1959–1981*, Sales and Publications, University of Strathclyde, 1994, page 118.
14. At Jordanhill the fact that the intake to Dunfermline College of Physical Education was always substantially higher was considered to be unfair, even although it was recognised that wastage rates from female members of the PE profession, but not from courses, was higher.

15. National Audit Office, *Scottish Education Department: The Control of Further Education Establishments*, Report by the Comptroller and Auditor General, HMSO, London, 5 June 1985, Appendix 3.
16. Ibid, paragraph 1.28.
17. Ibid, Appendix 4.
18. Ibid, paragraph 1.32.
19. David Semple, former Director of Education for Lothian, at a conference on the Implementation of the Munn and Dunning Programmes in Edinburgh, 15 October 1979.
20. Olga Wojtas, *Times Higher Education Supplement*, 16 January 1981, page 2.
21. Scottish Education Department, *Future Strategy for Higher Education in Scotland*, (STEAC Report), HMSO, Edinburgh, 1985, paragraph 5.28.
22. Ibid, paragraph 5.31.
23. STEAC Paper 85/63, submitted by T R Bone.
24. Letter to Mr McLeod from T R Bone dated 13 May 1985.
25. STEAC. Minute of meeting on 12 July 1985, paragraph 16.
26. STEAC Paper 85/65, Paper presented by H Cuming.
27. STEAC. Minute of meeting on 29 July 1985, paragraph 25.
28. STEAC. Minute of meeting on 12 July 1985, paragraph 18.
29. Ibid, paragraph 24.
30. Letter submitted by Cuming to the Secretary dated 10 September 1985.
31. STEAC. Minute of meeting on 29 July 1985, paragraph 22.
32. STEAC. Minute of meeting on 18 January 1985, paragraph 11.
33. STEAC. Minute of meeting on 12 August 1985, paragraph 9.
34. STEAC. Minute of meeting on 29 July 1985, paragraph 25.
35. STEAC. Minute of meeting on 12 July 1985, paragraph 21.
36. Ibid, paragraph 27.
37. STEAC. Minute of meeting on 29 July 1985, paragraph 25.
38. STEAC. Minute of meeting on 12 August 1985, paragraph 10.
39. STEAC. Minute of meeting on 12 July 1985, paragraph 18.
40. Ibid, paragraph 20.
41. Scottish Education Department, *Future Strategy for Higher Education in Scotland*, (STEAC Report), HMSO, Edinburgh, 1985, paragraph 5.34.
42. STEAC. Minute of meeting on 12 August 1985, paragraph 11.
43. *The Times Educational Supplement Scotland*, 15 February 1985.
44. "The unashamed professional", in *The Times Educational Supplement Scotland*, 18 February 1983, p 2.
45. Thomas Jones, "Who wants to be an unashamed professional?", in *The Times Educational Supplement Scotland*, 3 June 1983, p 2.
46. Gordon Kirk, "The BEd in Scotland: Towards a Professional Degree", Scottish Educational Review, Vol 16, No 1, May 1984, pp 19–26.
47. Walter M Humes, *The Leadership Class in Scottish Education*, John Donald Publishers Limited, Edinburgh, 1986, Chapter 7.
48. "The Future Structure of Higher Education in Scotland", *The Times Educational Supplement*, 20 January 1989.
49. Response of the National Union of Students to the STEAC Report.
50. Moray House submission to STEAC.
51. Gordon Kirk, "The Future Context of Professional Education" in *Moray House and Professional Education 1835–1985*, Scottish Academic Press, 1985, Chapter 15.
52. Gordon Kirk, Personal submission to the STEAC, September 1985.

53. Tom Bone, Paper submitted to the STEAC, 1985.
54. Statement by Secretary of State to House of Commons, 18 July 1986.
55. Scottish Education Department, *Report of Review of Teacher Training at the University of Stirling*, Edinburgh, 1988.
56. W Turmeau, *The Future Pattern of Higher Education in Scotland* (unpublished paper), 1988.

2

Associated College

Why linkage?

Relationships between institutions of higher education can take many forms, the spectrum of possibilities ranging from co-operation for a specific purpose to full integration or merger. In 1989 Moray House sought an association that was formal but restricted, and the term "linkage" was adopted as the most appropriate term to denote a mode of association which recognised the integrity of both institutions and was well short of full union. Merger was never considered and care was taken to ensure that the wording of the joint announcement made no reference to that form of institutional integration, though that precaution did not inhibit the subsequent press reporting. Moray House had no interest in surrendering its institutional autonomy and independence. Changes in the context of higher education, most obviously the frustration, through the demise of CNAA, of the aspiration for accredited status with, in due course, degree-awarding powers, made it imperative that an alternative validating authority was found. It was also assumed that the future well-being of the institution and the interests of its students and staff would be best served by seeking awards from a university. A formal linkage with a university would enable that objective to be achieved without any loss of institutional autonomy or any compromise of institutional mission.

Linkage with a university had two other attractions. Firstly, it would frustrate the apparent strategy of the Scottish Office to locate Moray House permanently in the non-university sector of the binary system it seemed determined to establish through the Conference of Centrally-funded Colleges. Secondly, it was believed that an association with a university institution could enable Moray House to enhance still further the quality of its work. Since 1981, Moray House had embarked on a major programme of course development and review with increasingly regular contact with CNAA. That contact resulted in a series of major achievements which brought about a transformation in the institution. However, still further development was required and it was believed that, if CNAA no longer provided a stimulus to institutional development, the interactions and collaboration which a linkage with a university would permit would enable the quality of the institution's work to be enriched and its standing enhanced.

Why Heriot-Watt?

The choice of Heriot-Watt as academic partner certainly startled the educational community in Scotland. The press reaction to the announcement was characterised by varying degrees of surprise and even astonishment. Musing on the announcement, the Editor of *The Times Educational Supplement Scotland* acknowledged that Moray House "had no historic links with Heriot-Watt, but rather a long-standing and frequently fractious relationship with the older university in the city".[1]

It is true, as Bell's (1983) study has shown,[2] that inter-institutional collaboration earlier this century in Edinburgh and elsewhere was characterised by squabbles of varying degrees of intensity. However, from the mid-60s universities and colleges of education entered a period of more fruitful co-operation. The BEd degree, which had been recommended by the Robbins Report in 1963, substantially increased opportunities for collaboration between university and college of education staff, and there is no doubt that the intensification of interaction generated by the degree was highly valued, as was the degree itself. Collaboration was also encouraged by involving college of education staff in teaching specialist modules on the MEd degree in a way that enriched provision for students at that level. It would also be fair to claim that relationships between staff in university departments of education and, particularly, the professional studies departments in the colleges were cordial, reflecting relationships which had been established while they were either students or teachers on the MEd. Finally, it was common for universities to nominate their Professor of Education to serve on the Board of Governors of a neighbouring college. For instance, in Edinburgh the Professor of Education, Noel Entwistle, served as a governor for many years, chairing one of the key sub-committees of the Board and working hard to support the college through extremely demanding times. That same helpful involvement was in evidence in other centres.

Despite that, there were several developments throughout the '80s which created tensions between colleges of education and university departments of education, if not the universities themselves. The decision by SED in 1982 to withdraw support for the BEd degree established in the mid-'60s following the Robbins Report was interpreted as an attack on a successful piece of inter-institutional collaboration that was valued by both staff and students. The case for withdrawing the degree was incontestable on economic grounds: in 1981 the BEd contributed only 5.4% to the output of qualified teachers and a reducing number of students had been distributed across a host of subject areas in six different colleges. Indeed, the SED consultation paper showed that over half of the 50 groupings across the system consisted of fewer than ten students.[3] Nevertheless, the links that had been established between colleges and neighbouring universities were very much valued, not least because they gave staff the opportunity of working to degree level. They also gave universities a foothold, albeit a tenuous one, in teacher education.

The degree which replaced the Robbins BEd – the BEd for Primary Teachers, introduced in 1983 – was clearly seen as an initiative that did not require collaboration with a neighbouring university and was interpreted as substantially reducing the role of the universities in teacher education. The fact that certain colleges, notably Moray House, sought validation for the new degree with the CNAA was interpretable as a gesture of ingratitude, setting aside almost two decades of fruitful collaboration on the old BEd. That interpretation was not strictly accurate. Once the overall shape of the Moray House new BEd had been agreed by the Board of Studies, there was discussion about the validation of the programme. The Board of Studies overwhelmingly argued for the degree to be validated by the University of Edinburgh, thus maintaining the tradition of collaboration with that university. However, the University of Edinburgh's ordinances and regulations did not permit it to act as a validating body for a programme to which it made no academic contribution and, despite appeals made at the Senate, the University indicated that it could not validate the degree. Moray House was therefore left with no alternative but to seek CNAA validation, an outcome that was not entirely displeasing to the Principal, given his strong involvement in CNAA work.

The announcement about the new degree created other difficulties. The introduction of a new award, even one that is long-awaited and warmly welcomed, always creates difficulties for two groups of students: those who have already or just completed the established programme, and those on-course during the years leading up to the introduction of the new award. The first group are usually accommodated by introducing an in-service route to the new award, with credit being granted for the former but superseded programme and sometimes also for professional experience as a teacher. Immediately after the introduction of the BEd, colleges of education introduced an in-service version of the award precisely to meet that need. The second group, those students already enrolled on the course that is to be superseded, are in a particularly difficult position and can feel desperately unlucky to miss out on the opportunity to obtain the new award. Conscious of this difficulty, and in response to pressure from students themselves, colleges sought to come to an arrangement with a neighbouring university to enable students, after a period of extended study beyond the three years of the diploma, to obtain a degree. Universities differed in the extent to which they could respond to this difficulty and at Moray House it was necessary to enter into negotiation with the University of Stirling, with which Moray House had been in association following the closure of Callendar Park College of Education in 1981, in order to obtain a degree award for those diploma students who wished to receive it. These negotiations did not get off to the best of starts, since the representatives of the University of Stirling opened the discussion by suggesting that Stirling might co-operate with Moray House as requested, on the understanding that Stirling, which did not have an intake to the new Primary BEd, would have an involvement in the Moray House BEd. It was not possible to accede to that request, but it was possible to make a degree of the University of Stirling available to students at Moray House who wished to extend their studies beyond the diploma and who, otherwise, might have remained bitterly disappointed and justifiably aggrieved.

A further deterioration in relationships occurred in 1985 when the SED withdrew financial support for Diploma in Education students. For many years, colleges and universities collaborated to the extent of allowing students taking the one-year postgraduate course to be exempted from courses in education and psychology at the college and to take, instead, corresponding courses at the university as part of a more extended programme of study which issued in the award of a Diploma in Education (DipEd). That was normally the first stage of the EdB (later MEd) Degree. Taking the DipEd was therefore an attractive option for students entering the postgraduate college course, for it gave them, in effect, a double qualification (the teaching qualification and the DipEd) and enabled them to embark on a programme of postgraduate study in education which, at that time, was almost a prerequisite for a post in educational management and other areas of the educational service. On the other hand, the DipEd was also seen as a way of reducing the exposure of graduates to life in colleges of education, which, according to undergraduate folklore, was intellectually sterile, was over-regimented and rule-ridden, and required graduates to rub shoulders with diploma students, an altogether inferior breed. Unfortunately, some leading members of university education departments, notably Walker of Aberdeen, Adams of St Andrews, and Pilley of Edinburgh, found ways of reinforcing that perception. They were the leaders of the campaign to remove the postgraduate course from colleges altogether, and managed to persuade the Robbins Committee to recommend that it should be transferred to the universities. From the colleges' perspective, the DipEd reinforced the perennial distinction between theory and practice and left them with those students who were judged not sufficiently able to be admitted to the DipEd course. For the universities, however, the DipEd symbolised their entitlement to participation in teacher education. The withdrawal of funds to support students taking the DipEd was therefore a devastating blow, confirming that the universities had no such entitlement or locus. For their part, the colleges of education could see in the termination of funding for the Diploma in Education an opportunity to achieve a more fully integrated course and therefore a strengthened postgraduate programme. The fact that staff in the colleges of education did not share common cause with their counterparts in the universities certainly soured relationships.

A particularly well documented demonstration of the tension between the two sets of institutions is provided by their respective involvement in the review of inter-institutional relationships in higher education conducted by the Scottish Tertiary Education Advisory Council in 1984. As was noted in Chapter 1, with the exception of Jordanhill College, which, throughout the review, held that the most appropriate development for the colleges of education was as faculties of education in neighbouring universities, the colleges of education were united in defence of the monotechnic principle: they were resolutely determined to maintain their independence. For their part, the university departments of education, writing collectively as the Scottish Universities' Council for Studies in Education, expressed themselves as forcibly in opposition to the monotechnic principle and urged the inclusion of teacher education within a broader institutional framework, such as that provided by the universities.[4] They were "sceptical about the desirability of the largely monotechnic nature of the colleges as at present organised" on the grounds that it reinforced "a narrow conception of professionalism confined to schooling and classroom competence", and concluded that "it would be a greater service to the educational service and to the colleges themselves if they were more closely and organically linked with universities or central institutions rather than being marginalised as at present". That statement may have reflected the marginal status of the university departments of education more than anything else. Its significance here is that it demonstrates how those in university departments of education could unite in demeaning the work of the colleges.

An even more public expression of the degree of ill-feeling then prevalent was occasioned by the review of SCOVACT in 1988, the body established as a collaborative initiative by the universities and colleges of education in 1983 and seen as offering an alternative validation service to CNAA. It came as something of a surprise to learn through the grapevine and not through any formal channel that the universities had decided to review SCOVACT. It seemed necessary to expose some of the thinking behind the review to public scrutiny and that was undertaken through the pages of *The Times Educational Supplement Scotland*. The Editor's choice of title – "A blatant pursuit of élitist self-interest" – certainly conveyed an appropriate degree of animosity. The article itself went on to describe the Trainer Report, as it was called, as "an astonishing document". It maintained that it was "totally inappropriate for one of the parties to the original agreement to the establishment of SCOVACT to assume, without consultation with the other party, an entitlement to initiate a review of the operation of the partnership and to enter into discussion with SED and others about the development of SCOVACT". The impression appeared to be given that the universities were concerned not to undertake an objective analysis of a complex question but "rather to lobby for a particular solution that will be to their own advantage". The article went on:

> *Given that the review body is composed entirely of university people and the limited conception of consultation to which the sponsors of the review appear to subscribe, what hope is there that the outcome will represent anything like a fair and dispassionate assessment? The whole arrangement is deplorable: it undermines the principle of professional partnership and collaboration; and it appears to expect colleges of education to assume an attitude of subservience, something that would be quite inappropriate, even if the universities were in a position to insist upon it.*[5]

Turning from the manner of the review to its content, the article identified "further grounds for disquiet":

> *The report turns out to be explicitly preoccupied with protecting the interests of the universities. There is no consideration of what is best for teachers; no analysis of the principles of quality assurance; and no consideration of the public and professional expectations of an external validation system. On the contrary, the report is an elaboration of how "the Scottish universities will undoubtedly stand to gain", of how their place in teacher education and professional development might be secured, and of the steps needed to protect the viability of their departments of education.*

Finally, exception was taken to the proposal that the university MEds should be officially endorsed by SCOVACT without being required to undergo external validation. That reference to postgraduate qualifications touched a nerve. For many years the Scottish Office had denied colleges of education the right to offer master's degree programmes, believing that that was a function of university education departments and effectively protecting the universities' monopoly, despite the fact that in several places college of education staff contributed to the teaching of MEd programmes. When, ultimately, colleges of education were able to compete, they soon attracted significant cohorts of students, at least partly because of their more favourable funding arrangements. It was a matter for concern, however, that, while college master's degree programmes had to be externally validated, there was no such requirement for university master's degrees. The universities' stance on this issue was promulgated chiefly by Nigel Grant, Professor of Education at Glasgow.[6] He had three lines of defence: that University MEds, having been approved by Senates acting under their statutory powers, were already "valid" and did not require the scrutiny of SCOVACT's validation procedures; that the £100 capitation fee required by SCOVACT would be passed by universities on to students and therefore would amount to "little more than a tax on the MEd"; and that, while there would be value in having university degrees "recognised" by SCOVACT as worthy of inclusion in Scotland's inservice provision for teachers, the precise nature of the process of "recognition" required to be clarified, it being claimed that "we are moving into uncharted waters: nobody has yet made it quite clear what the recognition of a degree as part of this structure will entail, what advantages will accrue to teachers taking these courses, or what financial and other support they can expect".

Each of these arguments was disputable. In the first place, the establishment of SCOVACT in no way threatened the autonomy of universities: degrees awarded following SCOVACT's validation were degrees of the individual universities, which had agreed to accept the judgement of SCOVACT on the academic value of a programme. No university viewed the external validation of the Primary BEd by SCOVACT as in any way "transgressing the established rights of universities".[7] Secondly, it did not say much for the universities' commitment to SCOVACT that they appeared unwilling to strengthen

SCOVACT's financial base by meeting the costs involved, preferring instead to pass these on to students themselves. Finally, the claim that no-one knew what was entailed by the process of "recognition" looked suspiciously like mere prevarication, a further rationalisation for refusing to subject university degrees for teachers to the same critical process which the universities required for other courses for teachers.

The Trainer Report devoted a significant amount of time to the MEd, pointing out that MEds developed by the colleges of education and validated by CNAA would "prove a further threat to the future of the university departments (of education)". Not surprisingly, having failed to take a college of education view on the matter, the Trainer Report came forward with the "recognition" argument: all that was required, in the case of the university degrees, was the kind of additional scrutiny that would determine whether or not these programmes should be "recognised" as part of a national structure of inservice awards for teachers.

The colleges of education were not impressed by semantics of that kind. The article on the review of SCOVACT found it "quite unacceptable" that programmes offered by the colleges would be subject to detailed scrutiny, while those from universities would "go through on the nod". And it considered that it was "simply nonsense to maintain to be committed to a national system of course provision on the one hand and, on the other, to seek a rationalisation for excluding some courses from the external validation process". Indeed, commentators were left to contemplate the undignified spectacle of universities establishing a review whose sole purpose appeared to be to secure a realignment of the goalposts.

The rejoinder to these criticisms was provided by Grant and Lyall,[8] the latter having served as a member of the Trainer Committee. Wondering what the Principal of Moray House was "getting so steamed up about", they claimed that the Trainer Report was an in-house affair, an attempt by the universities to establish a common position "with a view to discussions being undertaken with the colleges of education". It was not disclosed why this internal review entailed discussion with the Scottish Office, with the Scottish Committee for Staff Development in Education, the central body with responsibility for overseeing inservice provision, and with SCOVACT itself. These wider discussions, conducted without any consultation whatsoever with their so-called partners in the colleges of education, suggested that the universities were privately lobbying on behalf of their own position. Exception was taken to the use of the term "secret" to describe these discussions but there is no doubt that that adjective was appropriate: what took place was a private lobby on behalf of the universities. It appeared, indeed, that the principal aim was not to reform SCOVACT – that body was as good as a regional validating body was ever going to be, especially in view of the rather limited conception of quality assurance to which it subscribed – but simply to safeguard the university MEd. Exception was also taken to the "disgraceful suggestion" that SCOVACT practices could not be impartial, but that charge was also valid, since a conflict of interest was bound to arise whenever a university which offered an MEd was involved in the scrutiny of an MEd being proposed by a direct competitor.

This incident had the effect of convincing at least some colleges of education of the universities' bad faith, and further evidence of the attitude of disdain towards them which emanated from the universities, particularly from the Scottish Universities' Council for Studies in Education. Admittedly, the original critique of the SCOVACT review was intemperate: it was certainly meant to convey furious indignation. The rejoinder from Grant and Lyall, while purporting to be a contribution to "reasoned public debate", was little more than a polemic, spiced by personal insult and hyperbole, the most extravagant of which was that "as a long-standing enemy of SCOVACT and all its works, the Principal of Moray House would presumably not have been happy with any recommendation which fell short of the public execution of the convener of SCOVACT in the middle of the Canongate". So much for reasoned public debate! The episode provided further striking evidence of the gulf that existed between the colleges and the university departments of education, and of the degree of animosity that existed between them, hardly the basis for collaboration.

It would not be paranoid to claim that throughout all of these exchanges the Principal of Moray House was identified for special demonisation. He was seen as the most vocal defender of the independence of the colleges of education; he was an aggressive protagonist of CNAA validation and a belittler of SCOVACT; and he had presented himself as an enthusiastic advocate on behalf of the new BEd and the philosophy of professional education underpinning the new award. Not surprisingly, he was the focus of a good deal of critical comment by staff in the university departments of education, most notably in *The Leadership Class in Scottish Education*, in which Humes, then a member of the Education Department at the University of Glasgow, trenchantly attacked the notion of professionalism underlying the new BEd.[9] Whether or not the arguments advanced by Humes were valid, they were ineffectual and had no impact on the direction taken by the new BEd or on the work of the colleges. Indeed, in the context of the present discussion, Humes's analysis is interpretable as symptomatic of the division between university departments of education and the colleges.

His role as Chair of the Scottish Council for Research in Education at this time made the Principal of Moray House a target for even sharper attack. Appointed in 1984, at a time when the future of the organisation was in doubt and the complete withdrawal of SED funding was a real possibility, SCRE's new Chair developed with other Board members a draft strategy for the organisation's survival. An integral feature of that strategy was the proposal that SCRE should play a key role in research policy. That is, as well as conducting research and disseminating research findings, it was proposed that SCRE should have a role in allocating funds for research on the basis of priorities established through a process of consultation involving teachers, policy-makers and researchers. That was an attempt to democratise educational research and to place its funding on a more transparent basis, as well as to challenge the practice of allocating research funds on the basis of private discussions between researchers and SED officials. That feature of SCRE's future strategy was not popular with university departments of education, who drew attention to the danger that, as a research agency, SCRE might have a conflict of interest when it came to the allocation of research funds. Of course, there were obvious ways of avoiding that difficulty – for example, by stipulating that SCRE would receive the funds for its research from the Scottish Office and would have no entitlement to bid for funds from the pool set aside for competition. So deep-seated was the resentment at the SCRE draft strategy that when it was made public the university departments of education collectively submitted to the press critical comments of the measures that had been proposed to secure the survival of SCRE and were publicly critical of the confrontational stance that was apparently being adopted by the Chair.

The evidence is clear, then, that relationships between university departments of education and colleges of education were far from harmonious in the course of the '80s. Given the existence of these difficulties and rivalries, in which the Principal of Moray House, not one to eschew controversy, featured prominently, there were significant attractions, when it came to the point of deciding on a university partner, to turn away from a set of professional and institutional relationships that had not contributed to Moray House's development and from those who, collectively, appeared to be antagonistic to the work of the colleges of education in general and of Moray House in particular. Rather than reopen wounds and risk the resurgence of inter-institutional rivalries and inter-personal hostilities, it seemed sensible to turn to a new partner, an institution with which Moray House had no previous association. It was felt that Heriot-Watt would bring no baggage of resentment or distrust to discussions, and that a partnership could be developed with that institution without there being any question of settling old scores or pursuing an out-dated agenda. In short, the link with Heriot-Watt offered the opportunity for a fresh start.

There were, however, other attractions. Firstly, Heriot-Watt was a small university with a limited academic range. These features suggested that it would welcome an approach from an institution like Moray House. Secondly, there was virtually no academic overlap: the partnership would therefore be formed by institutions which were complementary. Thirdly, since there was no competition for students, difficulties over validation of courses were unlikely to arise, in sharp contrast to what might have happened if Moray House had gone to another university for the validation of a master's degree programme in a field of study in which that university already made provision. Fourthly, Heriot-Watt University had established a relationship with Edinburgh College of Art in which students obtained degrees of Heriot-Watt University while the institution retained its autonomy, its independence, and its distinctive mission. It seemed reasonable to suppose that what had proved acceptable at Edinburgh College of Art for some twenty years might also be appropriate at Moray House. Finally, the Heriot-Watt partnership, if it flourished, could provide a solution to one of the most intractable problems at Moray House – its accommodation.

Heriot-Watt was on the point of completing a major relocation to a greenfield site, one that offered significant scope for development. For its part, Moray House operated on three sites – at Newington, at Cramond, and at Holyrood. The Cramond buildings, dating from 1964, were nearing the end of their anticipated lifespan of twenty-five years. Newington was a continuing drain on the institution's resources, with summer letting, on which the college depended to make its student residences pay, falling away because of the inadequacy of the provision. And the Holyrood campus consisted of a mixture of buildings, some of them dating from the early 17th century, including the oldest of the college buildings in Scotland and some of the most hideous architecture in Edinburgh, which was widely acknowledged to constitute a blot on the landscape of the Old Town. Moray House's assessment of its accommodation and facilities at that time was that they were "in need of substantial enhancement". Some areas, such as the Library and the Cramond Campus, were either "extremely poor" or "seriously inadequate".[10] Too large a proportion of the college's estate funding was devoted to maintaining badly designed buildings.

For students and staff the inadequacy of the accommodation was a matter for continuing complaint and repeated student evaluations drew attention to the cramped accommodation that was either over-heated or under-heated, depending on the vagaries of an extremely wayward but expensively maintained heating system. Moreover, many of Moray House's regular visitors were distinctly unimpressed. One overseas visitor, who had come to inspect the college's computer facilities, questioned why such a substantial investment had been made in a slum, and a letter to *The Scotsman* referred to the famous summerhouse, set in the grounds of Moray House, as "a jewel in a pigsty". No doubt the condition of the buildings was partly attributable to faulty design and to sustained under-investment in maintenance over many years of severe financial pressure. Whatever the explanation, it was not right that staff and students had to tolerate buildings that were generally poor and, in places, seriously substandard. If an effective academic relationship could be established with Heriot-Watt, relocation to new facilities on a single site at Riccarton might solve a major problem which, left to itself and in the funding regime then prevalent, Moray House would have found it to be well beyond its capacity to address. That problem alone called for the kind of radical solution which the academic linkage with Heriot-Watt signalled.

Statement of Intent

Principals have an obligation to consider the strategic development of their institution; they need to attend to developments in the environment of higher education; and they must regularly test their reading of the signals suggested by policy and other changes with their senior colleagues. The decision to approach Heriot-Watt University grew out of discussions with the Chairman of the Board of Governors, John Cook, and the Vice-Principal, Edmund Ewan, early in 1989. Feelers were put out to confirm that an approach for discussion would receive a positive response and the first meeting took place at Riccarton in late May. Moray House was represented by the Principal, the Vice-Principal, and the Chairman of the Board of Governors, and Heriot-Watt by the Principal, the Vice-Principal, the Deputy Principal, and the Secretary. It took roughly half-an-hour to reach agreement that a formal association should be explored, Moray House having acknowledged that, naturally, it would need to switch from validation by CNAA to Heriot-Watt's course approval arrangements. It would be exaggerating to say that this was any kind of concession, since it was obvious that there would have been no point in a meeting at all if there had not been an intention to secure Heriot-Watt degrees for Moray House students. The meeting ended with the agreement that a statement setting out a number of general principles should be prepared for discussion at a future meeting. Moray House undertook to draft that statement.

The Statement of Intent, as it came to be known, was drafted with the needs of a number of audiences in mind. Since a transbinary initiative involving a teacher education institution was bound to create some eyebrow-raising in the Scottish Office, it was essential to have regard to the position of the Secretary of State and to the prevailing assumptions of politicians and others in officialdom. The Statement of Intent was therefore drafted with an eye on Michael Forsyth's speech to Chairmen and Principals of Centrally-funded Colleges at Dunblane on 8 May. Clause 2 could have been written by the drafter of that speech: it affirmed the belief of both institutions that "through the planned use of shared resources they can in partnership enhance the quality of the higher education they provide, extend opportunities for higher education and continuing professional development, and offer a more cost-effective service to the community". Subsequent references to "effectiveness, efficiency and economy" invoked the prevailing mantra of politicians and others. Besides, the reference to "the shared commitment to excellence in vocational higher education" sought to allay ministerial fears of academic drift; and the stipulation that quality assurance procedures should have "a strong external involvement" was likely to reassure those in the Scottish Office who were strongly committed to external validation, even for universities, and who might see in this clause the thin end of a highly convenient wedge. As if these various attempts to assuage official suspicions were not enough, the statement made three specific references to the need to protect the position of the Secretary of

State (and also of the General Teaching Council) with regard to teacher education courses by acknowledging that an association between the two institutions "will require the approval of the Secretary of State for Scotland". It was for that reason that the statement was discussed with SED officials on a "without prejudice" basis before it was made public.

The Statement of Intent also sought to secure proper protection for the Moray House community. To that end, reference was made to "preserving the institutional identity of Moray House College"; a commitment was made to secure Heriot-Watt degrees for students enrolling in 1990; and an explicit undertaking was given which protected the conditions of service of staff. Furthermore, it was agreed that no decision would be taken on the outcome of the proposed Joint Working Party "without full consultation within each institution", and the development of the initiative would "only be on the basis of agreement between the two institutions". Finally, clause 5 of the statement opened up the possibility of relocating Moray House to the Riccarton campus, a clause which invited sceptics at Riccarton to contemplate the size of the dowry accruing from the disposal of Moray House's three campuses. Care was taken, however, to ensure that the provision of Heriot-Watt degrees for Moray House students was not contingent on the outcome of the Working Party report and would take place in advance of any decision on relocation, thus removing the possibility that Moray House would be obliged to trade its premises for Heriot-Watt degrees and effectively keeping Moray House's options open.

Having been agreed by senior officers of both institutions, the Statement of Intent was discussed at a meeting of the Chairman's Committee of the Moray House Board of Governors on 23 June and agreed a few days later by the full Board. It was published on 28 June 1989 and was the focus of widespread media interest. According to the press of the following day, the reactions of various members of the educational community were marked by "surprise" (*The Scotsman*), "astonishment" (*The Times Educational Supplement*), and "anger and suspicion" (*The Daily Record*). Education unions seemed to be particularly aggrieved and were variously described as "shocked and surprised" (*The Glasgow Herald*) and "stunned" (*The Times Educational Supplement Scotland*). *The Scotsman* sought to explain the surprise on the grounds that "this was a match that nobody seemed to have predicted", and by suggesting that "until yesterday, the Head of Moray House was known throughout Scottish educational circles for his fierce condemnation of mergers". One union official, Bob Beattie of the EIS, thought it dangerous that Moray House might be "submerged within a greater campus" (*Edinburgh Evening News*) and maintained, "Moray House is currently in the centre of a great city. If this merger goes ahead the college will not be as accessible to many teachers who rely on it for in-service training and other services."

Ian MacPherson of EIS/ALCES, the academic staff union, expressed "total condemnation of the college authority's action" and maintained that the proposals "should be abandoned if they failed to secure the support of all staff and students".[11] Jack Dale, the Further and Higher Education Secretary of the EIS, was particularly exasperated. He maintained that "this free-for-all restructuring of higher education in Scotland must be brought under some kind of control", a reference to the fact that in recent months merger discussions had been intimated involving several Scottish higher education institutions, including Heriot-Watt University and the Scottish College of Textiles.[12] He argued that "colleges are not private companies to be traded in the market place, and were not at the personal disposal of their principals".[13] He and other union officials were particularly concerned because staff at Moray House were told of the talks less than an hour before the press release. That was taken as evidence of a lack of consultation. Proposing the establishment of a top-level national committee to scrutinise proposals in the interests of maintaining a coherent system of higher education, he concluded that "the system of higher education is being restructured by default". More than anything, he deplored such "back-door methods" of developing institutional associations.

While it was gratifying to note that the significance of the initiative had been recognised by the press by such observations as "Moray House first across the binary line"[14], the general press reaction was disappointing. Four observations might be made. First, the plea for a national consideration of the structure of higher education was an attempt to stem the tide. Michael Forsyth, in line with his political principles, had moved the debate beyond corporatism: it was now imperative for institutions to work out their own salvation. Moray House had faced that responsibility and it had judged that its future lay in collaboration: other institutions would not be far behind.

Secondly, the charge about lack of consultation was also wide of the mark. Of course, principals require to consult with their staffs, and consultation was deeply embedded in the Moray House culture. However, principals have a role as academic leaders: they have an obligation to make proposals on the way forward. The Statement of Intent consisted of proposals precisely of this kind. They may have been more fully elaborated than is common in such circumstances, when principals simply disclose that talks are under way, a strategy that can often lead to even more uncertainty for colleagues who wish to know what the talks are about. The Statement of Intent sought to disclose what the substance of the talks had been and to reveal where the principals' preferences lay. The document was clearly consultative and fully acknowledged the need for further discussion. It was therefore erroneous to interpret the Moray House approach as evidence of a lack of commitment to consultation, as one principal of a rival institution implied. That was based simply on a failure to read the text of the document and may have been based on other reasons for seeking to discredit the initiative. Nor was it a matter, as Willis Pickard claimed, of simply "letting the staff into a secret".[15] In any major initiative of this kind, particularly one that proves to be politically sensitive or to challenge received assumptions about the structure of higher education in Scotland, a reasonable degree of confidentiality is essential to test the extent to which there appears to be agreement at senior level in both institutions. As it was, the opposition to the proposal that the press claimed to detect was not borne out by events: the subsequent meetings that took place with union officials were perfectly cordial and provided all the assurances that were sought, including full participation in the subsequent consultation process.

Thirdly, there was no question of a *fait accompli*. The Statement of Intent clearly acknowledged that the validity of the proposals required to be tested and that much more detailed work had to be undertaken before a final decision was taken. In the press reaction it did not seem to be acknowledged that the proposals were entirely provisional and under-developed, and that, before they could take the form of a definite line of institutional development, the governing bodies of both institutions would need to agree, and the staff and students concerned would need to register their support. Furthermore, even if agreement was reached within both institutions, the Secretary of State would require to be persuaded of the validity and acceptability of the proposal. Given the need to secure agreement at these various levels, it is surprising that the media and the unions should conclude that a merger of the institutions was inevitable. Only Willis Pickard of *The Times Educational Supplement Scotland* appeared to appreciate the provisionality of the Statement of Intent. He wrote:

> *Free from the carefully documented minutes of civil servants, the principals of the two institutions and their immediate advisers were able to start and finish their initial talks in little over a month. Of course, matters are not at an end. It is only the beginning of the end. Genuinely, no-one seems to know whether ultimately there will be a merger or some association short of that. . . . The media are anxious to prove "mergermania". That is at best premature and inherently improbable.*[16]

That was an accurate assessment. It was also unique, for the rest of the press accepted that merger was the inevitable outcome of the announcement. It was repeatedly emphasised in the press conference and subsequent discussions that there was no question of a merger. Indeed, the Principal of Heriot-Watt University made repeated reference to a multi-collegiate University and went out of his way to emphasise the importance of the Moray House tradition, giving an assurance that the Moray House title would be maintained in view of its "honourable tradition and unique marketing value".[17] It is probable that the emphasis on merger was influenced by the over-elaboration of the matrimonial metaphor. That was stimulated by Professor MacFarlane's statement at the press conference that, "This engagement between Heriot-Watt and Moray House is a happy one. I am sure a marriage can be arranged." That may have been a latent intention at Heriot-Watt, but it was simply not on the Moray House agenda.

Proposals for Linkage

In line with the Statement of Intent, a Joint Working Party representative of both institutions was established and met for the first time on 22 August 1989. At that meeting a number of joint sub-committees were formed to consider academic matters, the constitutional relationship between the two institutions, finance, and buildings and estates. The aim was to generate a report by the end of December 1989. However, significant difficulties arose, particularly on the ways in which the institutional identity of Moray House would be protected. Several one-to-one discussions took place between the two principals on this matter. That caused a delay in the development of the Joint Working Party report, which did not appear until late March 1990.

The consultation paper *Proposals for Linkage* offered a vision of education for the 21st century in which the resources of technology were harnessed to the aim of extending educational opportunities for all in "innovative, flexible and supportive learning systems". It proposed a fusion of technological and educational skills in a partnership that would "play a leading role in the development and application of the new educational techniques and methodologies which will underpin new forms of education and training at all levels". The benefits of institutional association were adumbrated. For Moray House these were thought to lie in the further strengthening of its courses, in enhancing the quality of the students' educational experience, in achieving membership of a university community, and in the possible relocation of the college. From the point of view of Heriot-Watt University, it was claimed that a linkage would "serve to broaden the academic scope of the University"; that it would benefit by "establishing a formal association with an institution whose raison d'être is the fostering of human resourcefulness through learning"; that it would provide a more appropriate ratio of males to females in the University; and, finally, if it led to the relocation of Moray House, the linkage would entail a more effective use of "an expansive and attractive campus". Both institutions, it was maintained, would stand to gain by together making a major advance in the application of technology to teaching and by ensuring that the education of future teachers took place in a university environment, especially a technological environment, and one that equipped students with a richer understanding of the significance of industry, commerce and technology for the national well-being.

Several criteria were then identified for the evaluation of an appropriate institutional association. These were: enhancement of the quality of work; external scrutiny of courses; partnership with the professions; professional orientation of courses; improved effectiveness, efficiency, and economy; preservation of the Moray House identity; and the authority of the Senate and of the Secretary of State being unimpaired.

The central function of the proposed academic collaboration concerned the validation and review of Moray House programmes. For that purpose, it was proposed that an Academic Advisory Council should be established "with a majority of external specialists drawn from other institutions and from the professions on a UK basis" and with the following functions:

1. to review the quality assurance and quality control mechanisms and procedures used by the college;
2. to undertake periodic appraisals of the work of part or all of the college;
3. to preserve the external scrutiny of college courses.

Two forms of association were envisaged. In Model 1, the Moray House Board of Governors would be retained and its Academic Board would be responsible to the Senate in the same way as a faculty board; members of Moray House staff would become members of the Senate with reciprocal University membership of the College Academic Board; there would be University involvement in academic staff appointments and promotions; and students would become matriculated students of the University and would receive awards of the University. In Model 2, the University "would assume the responsibilities of Moray House College Board of Governors"; all the academic features of Model 1 would be retained but the staff would become employees of the University; and funding for the college would be made by the University Court with the Principal of the University as the Accounting Officer. In other words, Moray House would become, in effect, a constituent college of the University, linked to the Senate formally through its Academic Board, while all the responsibilities of the Moray House Board would be transferred to the Court.

A significant section of the consultation paper was devoted to the relocation of the college, setting out the benefits that would accrue from working in a new, purpose-built and more attractive physical environment. Naturally, the counter-arguments were also emphasised. Reference was made to the desirability of retaining a city centre location from the point of view of its attractiveness to students and the provision of in-service courses, and to the possible impact of relocation on staffing, both current and future. Having considered the arguments, the Working Party recommended in favour of relocation. The central recommendation was that a Model 1 form of association should be established from the beginning of session 1991/92 and that a financial appraisal be undertaken of the relocation of the college to Riccarton. It was also recommended that both the Board of Governors and the Court should be requested to have the paper widely distributed inside and outside both institutions to test the acceptability of the proposals.

At Moray House all members of the college community participated in the most wide-ranging consultation exercise ever conducted in the institution. Undoubtedly, the consultation period was dominated by the relocation issue. While it was recognised that the full benefits of the academic association could only be fully realised through relocation, serious reservations were expressed about the impact of relocation on the college's capacity to attract students, on job security, and the institution's attractiveness to teachers on in-service courses.

Summarising the institutional response, a paper from the Principal to the meeting of the Board of Governors on 26 June 1990 stated:

> *There is clearly widespread support throughout the college for the establishment of an academic linkage with Heriot-Watt University. The academic and other advantages of collaboration set out in* Proposals for Linkage *have been clearly acknowledged and there is agreement that the enhancement of the work of the college is to be secured by locating its activities in a university context.*

That paper went on to recommend that a Model 1 form of association should be adopted without delay; that the significant institutional advance which Model 1 could confer should not be missed because the major issues raised by Model 2 took too long to be resolved; and that, since Model 2 had some support, further studies should be undertaken and completed by the end of the following session. These would include an independent financial appraisal of the feasibility of relocation, as well as a survey of the views of students and practising professionals on the possible impact of a change of location.

Twenty-six institutions, most of them education authorities and other institutions of higher education, submitted observations on *Proposals for Linkage*. All of these were supportive or strongly positive, with education authorities expressing themselves in favour of the proposal but wishing to emphasise the need for the college to continue to be responsive to local needs.

Since the consultation process inside and outside the institution showed massive support for the initiative, the Board of Governors, noting in the report from the Principal that "there is a clear institutional consensus on the creation of an association between the college and Heriot-Watt University", agreed to the recommendations in the Principal's paper. Consequently, on 26 June 1990 the Principal was able to write to the Scottish Office indicating that the governors wished to proceed to an association with Heriot-Watt University under Model 1, subject to the agreement of the Secretary of State. The target set was the beginning of session 1991/92.

Subsequent correspondence with the Scottish Office was concerned with the validation of courses. It was known that "the Secretary of State attaches high importance to the concept of external validation of teacher training courses" and it was reasonable to expect the SED to probe the robustness of the course validation proposals. It was also questioned why the prevailing practice of using SCOVACT to validate college courses was not followed in this instance, "particularly as Heriot-Watt was a signatory to the SCOVACT agreement". There is evidence here of some lobbying on the part of SCOVACT. On

30 April the Secretary to SCOVACT had written to Heriot-Watt University drawing attention to the mandatory condition in the Primary BEd guidelines, which required that "the course will be validated by an external body", reminding the University that the Executive Committee of the Scottish Universities had only recently confirmed its support for SCOVACT, and repeating his offer to provide facilities for external validation of courses taught at Moray House. These points had been put in a meeting with the Principal of Moray House on 17 October 1989 and had received a dusty answer. The college replied fully on 6 September to the SED, strongly affirming the institution's commitment to external validation and drawing the Department's attention to the fact that, over the years, CNAA external validation had been weakened to the extent that the degree of externality had been reduced in what was now known as joint validation, and setting out the three grounds which made SCOVACT inappropriate as a validating body. First, through the CNAA, the college had become used to drawing on a national network of expertise and, in that way, being able to satisfy itself that its work was on a par with the best academic and professional practice in the UK. Secondly, the college had benefited from the CNAA institutional review procedures whereby its whole mode of operation was subject to critical scrutiny, a facility not available under SCOVACT. Thirdly, Moray House offered a number of programmes outwith the area of teacher education and was committed to the expansion in the number of such programmes. SCOVACT was clearly not an appropriate validating body for such programmes and there were advantages in having all of the institution's work validated by the same authority. Finally, the reply fully reassured the Department that the Secretary of State's position would not be affected in any way by the linkage.

On 18 February 1991 it was possible to write to governors indicating that the Secretary of State had agreed to the proposal that an academic linkage should be formed with Heriot-Watt University, as the Board had requested. It was then necessary to proceed to form a legal agreement with the University in the form of an Ordinance which set out the interim arrangements for students currently enrolled on CNAA validated courses and which gave students who had enrolled prior to 1991/92 the option of graduating with a CNAA award or an award of Heriot-Watt University. The overwhelming majority opted for the second of these. Arrangements were therefore in place for Moray House to proceed as an associate college of Heriot-Watt University from the beginning of session 1991/92 as planned.

Towards a More Integrated University

From the beginning of the session in which the formal linkage operated, the Principal and Vice-Principal of the University and Deans of the Riccarton faculties met with the Principal and Vice-Principal of each of the three associated colleges, the other two being Edinburgh College of Art and the Scottish College of Textiles. In the main, these meetings were concerned with the processing of Senate business, but they also provided the opportunity to consider academic developments and to monitor the relationship between Heriot-Watt University and its associated colleges. Inevitably, since major changes in higher education were under way, including legislation, the discussion turned to the ways in which the University and its associated colleges would need to accommodate and, if possible, exploit, these changes in the years ahead. The group, which came to be known as the University Joint Council, came to sense that in the intensely competitive environment that was being created in higher education there would be many areas in which the interests of all four institutions coalesced. Accordingly, the discussions considered how the four institutions might become more closely integrated, to create a single university framework within which the diverse fields of study would flourish.

At the meeting of the University Joint Council in May of 1992 a paper was brought forward by the University bearing the title *Heriot-Watt University Group*. It proposed "a gateway for single-source funding" from SHEFC, and the integration of the enlarged university with a single Court, a single Senate, a single Board of Management in each college exercising certain powers delegated by the Court, with representatives of each college Board of Management on the Court and on the Senate, and with standard university terms and conditions of employment for all staff. What was proposed, in effect, was merger. After discussion it was agreed that the four principals would develop the paper. While the Heriot-Watt paper had interesting features, it re-awakened fears that had been expressed frequently during the discussions prior to the establishment of the formal linkage. The central question concerned institutional autonomy. It seemed important to ensure that a decentralised approach to the merged institution would be adopted and that argument was advanced in a letter to Heriot-Watt. It maintained:

> *While the Court is the supreme decision-making body in the institution, it would be inappropriate if, given the fact that the institutions are geographically dispersed, there was not significant delegation of responsibility and function. We propose, therefore, that for each of the constituent colleges there should be a "sub-court" which would have a number of responsibilities relating to the college. The sub-courts would have no responsibility for negotiating pay and conditions, which would be the responsibility of the Court.*[18]

Turning to academic governance, the letter proposed that, while the Senate should continue to be the supreme academic decision-making body, it was not appropriate for the Senate to be responsible for all the academic decision-making in the institution. It argued that, "There is a strong case for the delegation of academic responsibility: the Senate that seeks to approve everything risks becoming a mere rubber-stamping body. Its function should be the determination of academic policy and the monitoring of the academic performance of the institution."

Following further discussion, the amended paper entitled *Towards a More Integrated University* appeared on 26 June 1992. The letter from Moray House congratulating Principal MacFarlane on its appearance described the paper as "a courageous and forward-looking document". Having rehearsed the arguments for closer integration the paper adduced a number of advantages, including breadth of provision in vocational and professionally-related courses and economies of scale, that would follow from closer collaboration between the four institutions. The combined advantages were thought to constitute a significant educational resource in the east of Scotland.

The crucial feature of the paper concerned the governance of the integrated university. What was proposed was a federal structure with a single university court and a single senate, on each of which the four institutions were represented proportionately, with substantial functions delegated to the four constituent institutions.

It was agreed that the paper, which had been confidential to the four principals, would be discussed at a dinner to which the Chairs of the governing bodies would be invited, to be held on Wednesday, 23 September 1992. It was not a happy meeting. In his words of welcome, the Chair of the University Court referred to the paper developed by the four principals and intimated, to the astonishment of everyone, that the paper "was totally unacceptable to the Court of Heriot-Watt University". The principals were charged to revise the paper and what emerged, after weeks of further discussion and negotiation, was a totally emasculated version, with all reference to federal arrangements eliminated. Some concession was given to the principle of delegation, but the original proposals on the composition of Court and Senate were withdrawn, as a way of affirming the dominance of Heriot-Watt University.

A special meeting of the Moray House Board of Governors was held on Tuesday, 3 November 1992. At that meeting it was striking how strongly members felt that "nothing should be done which damages Moray House's international reputation as a centre of excellence in professional education".[19] Nevertheless, it was unanimously agreed that the paper should be widely distributed and the results of that consultation process should be presented to a further special meeting to be held on 29 January 1993.

For its special meeting on 29 January 1993 the Board of Governors had all of the responses to the consultation exercise, with a brief covering paper from the Principal. That paper identified a number of options, with a preference for intimating to Heriot-Watt that "certain essential pre-conditions had to be met before any further discussion took place".[20] After extended discussion, the Board unanimously agreed the following motion:

> *The Board is unwilling to agree to move towards a fuller form of integration as set out in the discussion paper prepared by the four principals*

> *but, nevertheless, would wish to continue to build on the academic collaboration already under way. The Board would encourage the creation of a new form of institution, one based on the coming together of four equal partners, with arrangements to preserve the identity and maximum autonomy of each, and would welcome the University's formal confirmation that it shares that vision of future collaboration as the basis for further discussion.*[21]

Privately, the Principal of Moray House wrote to the Principal of Heriot-Watt in the following terms, offering his own personal observation:

> *I really do believe that the only way forward now is for the University Senate and Court to accept the principle that we are four equal partners, committed to the creation of a new federal university, as opposed to merely the incorporation of existing colleges within an enlarged Heriot-Watt University, and I am bound to say that, unless the University is able to make a public affirmation of that kind, I, personally, see no point in any further discussion between us on this matter.*

Meanwhile, the integration paper had been discussed by Heriot-Watt University Court on 16 November 1992. That meeting "prescribed a general framework within which any merger negotiation should proceed". That involved the following principles:

(a) the University must be convinced that any merger would enhance the University;
(b) the University must not lose its own identity or its reputation;
(c) the associated colleges would be required to merge by incorporation into the existing University;
(d) there must be no dilution of the University's standards;
(e) the importance of quality research and teaching must be recognised and actively promoted.[22]

The Court also prescribed the constitutional framework within which any merger negotiations should proceed. That would involve a single Court, a single Senate, and a single administration, but with sufficient changes to the University Charter and Statutes to enable the identities of the associated colleges to be maintained.

On 3 March, the Principal of the University wrote to report the reaction of the Court of the University to the submissions from the three associated colleges in the following terms:

> *After a full discussion of the principles and constitutional framework against which any further discussion might take place, the Court*

> *agreed to note the views of the governors of Moray House College and to confirm that the present collaborative arrangements would continue for the time being. The college's proposal for the creation of a new form of institution based on the coming together of four equal partners was considered to be incompatible with the principles agreed by the Court to form the basis of any further discussions.*[23]

With these words the prospect of an exciting new development in higher education in Scotland, one that would have secured a central role for Heriot-Watt as the hub of a network of federated institutions, perhaps known as the Federal University of Scotland, was finally eclipsed.

The Search for an Accreditation Agreement

The rejection of the integration paper brought a protracted round of formal negotiations on governance to a close, at least temporarily. However, the tensions inherent in an ambiguous institutional relationship, in which three independent institutions sought to protect their autonomy while professing membership of the same university community, was bound to be reflected in misunderstandings and disagreements. Besides, the constitutional issue was so central that it was bound to re-surface. Discussion was resumed early in February 1995. The context for the discussion was the impending visit by the Higher Education Quality Council (HEQC) to scrutinise Heriot-Watt's collaborative arrangements with its associated colleges. It was judged that it would be necessary to resolve the constitutional question before formal discussions began with the HEQC auditors. In anticipation of that visit, a meeting of Principals and Vice-Principals of the three associated colleges was held in Edinburgh College of Art. The discussion ranged over the causes of our prevailing discontent: unacknowledged correspondence; delays in the appointment of readers and professors; the inappropriateness of the criteria for promotion; the culture of rubber-stamping that seemed to characterise the Senate; and, most serious of all, the intimation, without any consultation, that representatives of the associated colleges would no longer serve as members of the Teaching Committee or the Research Committee of the University. Reflecting the mood of the meeting, Professor Rowan of the Art College wrote to the University in these terms:

> *I think it is fair to state that there is a general concern that the forthcoming visit by the HEQC Audit Team to the University may not be able to report positively on our existing institutional arrangements and that fundamental issues and concerns, which may be familiar to you and your senior colleagues, need to be aired and discussed in advance of our individual meetings as Principals with the Audit Team.*[24]

In anticipation of a meeting with the University, a paper was drawn up, drafted by Moray House and agreed by all three colleges, proposing a change in the constitutional relationship with the University. It began with these words:

> *All three colleges wish to continue as associated colleges of Heriot-Watt University with full accredited status. Indeed, we think that we should proceed on the assumption that the relationship with Heriot-Watt University is a permanent one.*[25]

It went on to propose a form of accreditation based on the CNAA Articles of Accreditation, whereby an institution is entrusted with responsibility for ensuring that academic standards are protected. The paper was based on the assumption that all three associated colleges had well developed quality assurance systems and all of them were well capable, with the multiple external points of reference they had established, to ensure that standards were maintained. It fully acknowledged the right of the University to scrutinise the quality assurance procedures of the three colleges from time to time but gave full responsibility to the colleges to approve courses leading to Heriot-Watt University awards. Provision was made for the regular auditing of each college's procedures at intervals of not more than five years.[26]

That letter was sent to Heriot-Watt University on 3 March in the expectation that the paper would be considered at the meeting of the University Joint Council scheduled for 7 March. Unfortunately, no discussion took place with the University prior to the visit of the HEQC later that month. Informally, the three Principals of the associated colleges agreed to take the line in discussion with HEQC that, while there were undoubted difficulties, discussions were taking place as to how these might be resolved.

The report of the HEQC visit was not a flattering one. While it was sent to the University on 5 May, it was not shared with the associated colleges until mid-June and, even then, only after the strongest of representations made by the associated colleges. The explanation for the reluctance to share the contents of the report was that "it was confidential to the University".[27] However, a more plausible reason for the secretive approach to the document was, as the University itself was forced to recognise, its extremely critical review of the University's arrangements for quality assurance and quality control. Inconsistencies were detected in the arrangements for all three of the colleges; the Academic Audit Group which had been given responsibility for the quality assurance arrangements in the three colleges "had not undertaken any action"; there was "no explicitly stated policy for reviewing the procedures by which the University assures itself of the quality of continuing collaborative programmes"; the University had "no formal mechanism by which it can assure itself of the quality of teaching and student learning in collaborative provision", other than that implied by external examiners; the University had "no formal mechanisms in place to assess the needs or efficacy of the student services accorded to college students", other than those provided by the Careers and

Library Service; the University had no record of the grades of entrants to its collaborative programmes and therefore was unable to make comparisons between entry and awards; despite the fact that the collaborative arrangements had been in place for several years, "little progress appears to have been made to ensure equivalence of award classifications across the University and its associated colleges"; and the University had not appeared to have grasped the distinction between annual monitoring and periodic review.

Perhaps the most significant feature of the report was the evidence found of straight malcommunication between the University and its associated colleges. For example, the quality audit report of Heriot-Watt in 1993 had referred to the University's intention to develop a manual of good practice. Two years later it was "still in preparation", despite the fact that two of the associated colleges had academic standards handbooks; while matters pertaining to the colleges were discussed at the Principal and Dean's Committee, the representatives of the colleges were not members of that committee; the University had discontinued college representation on key University committees; the University had not shared the results of its own Quality Audit Report of 1993 with the associated colleges; and there was little evidence of collaboration between the University's Institute for Computer-based Learning and the Scottish Interactive Technology Centre based at Moray House.

These observations contrasted sharply with those of the Quality Audit Report of Moray House, admittedly a year later, in July 1996:

> *In its sampling of the Institute's process the Audit Team found highly motivated and enthusiastic students, a vibrant staff, and a well managed institution. Throughout the audit, the team encountered abundant evidence of the dedication of members of the teaching and support staff of the Institute, led by its senior managers, to sustaining a rich and supportive environment to the benefit of its students. The Institute understands its role in professional teaching and learning in the Scottish context and beyond. It is located in an environment where it is held in high regard by the schools with which it works, and is part of a network of educational provision that is well respected and understood. Overall, the Audit Team found much to applaud in the enthusiasm and vitality of the staff and their commitment to an ethic of good professional practice underpinned by sound training methods, founded upon the distinguished tradition of teacher education in Scotland.*[28]

And, again:

> *The Team had a sense that the ethos and approach of the Institute differed from the University, such that effective communication and integration was difficult and, as such, impeded the enhancement of quality.*

The University clearly took the report of the HEQC very seriously. The Court meeting on 26 June 1995 was informed that the recent audit of collaborative arrangements "had been very critical of the present situation". The Executive and Deans' Committee was charged to prepare a package of measures for consideration at the meeting of Court on 25 September. That meeting of the Court approved a draft ordinance on accreditation, which was obviously judged by the University to constitute a comprehensive response to the difficulties that had been identified by the HEQC. On 8 November Heriot-Watt wrote to the associated colleges informing them of the Court's response and of the decision to engage in discussions with the colleges individually and in sequence, beginning with Edinburgh College of Art, then the Scottish College of Textiles, followed by Moray House. The letter also made it clear that, in addressing the issues, "it will be necessary to determine the effective action required to meet the adverse comments of HEQC following its audit of current collaborative arrangements".

The draft ordinance on accreditation was extremely disappointing. It proposed the establishment of an Accreditation Board whose composition would be determined by the University and would include six external members. In what was clearly an arrangement for validation rather than for accreditation, the draft indicated that it would be the responsibility of the Accreditation Board, which excluded from its membership the Principal and Vice-Principal of the college, to advise the Senate on quality assurance and to monitor quality mechanisms; to undertake periodic appraisals of all or part of the work of the institution; and to "advise the Senate on the approval of courses of study and on matters relating to course structures and syllabuses for University awards". It was envisaged that these arrangements would extend over five years, at the end of which, subject to review, full accreditation would be granted.

Growing Disaffection

Against a background of growing disaffection, some disagreements erupted into conflict. The University's handling of promotion to reader and professor strikingly exemplified how ferociously acrimonious that conflict could become. At its meeting on 12 September 1995 the Moray House Board of Governors received a paper from the Principal setting out the background to the use of these titles. In line with the agreement with the University, and also with practice in other parts of the University, appointments at Moray House to reader and professor were made by a University committee, the Senior Academic Staff Promotion Board. The paper indicated that the arrangements were not operating effectively, mainly because, in the University's established procedure, there was "a predominant criterion: the number of articles in refereed journals". It was acknowledged that that was a widely accepted indicator of academic distinction and was the criterion that was used in the promotion of other readers and professors on the Riccarton campus. However, its application to Moray House staff was judged to be inappropriate for three reasons:

(a) much of the research and development work of Moray House staff took the form of participation or playing a leading part in national projects which are published as official SED reports rather than articles in journals;
(b) other development work took the form of consultancy with practising professionals and there the indicator of distinction was not so much published work as its impact on professional practice;
(c) no provision was made for excellence as a teacher, notwithstanding the fact that the Principal of the University had chaired a national committee which recommended that excellence in teaching should be made a criterion for promotion to professor.[29]

In the light of these criticisms, the paper from the Principal proposed that Moray House should proceed to make promotions on its own, without reference to the University. However, before proceeding, the Board considered that it would be prudent to receive the views of Heriot-Watt University on that proposal. Accordingly, the Principal wrote to the Secretary of the University to obtain the University's view. That letter made it clear that, while the Board would prefer to proceed on this matter with the full agreement of the University, if that agreement could not be secured the Board was minded to introduce its own arrangements for promotion to reader and professor before the end of the term.[30] A reply to that letter was received from the Secretary of the University making it very clear that the University would be totally opposed to any such arrangement and that, if Moray House were to proceed, then the position of existing postholders of reader or professor would be put at risk. The letter expressed itself strongly against "a two-tier system of readerships and professorships".[31]

Moray House's reply to that letter aggressively repudiated the arguments advanced by Heriot-Watt. It was claimed that the criteria for promotion were ambiguous; that the arrangements gave insufficient weight to the judgement of people who had relevant academic expertise and credibility; and that "the criteria are inimical to the fair recognition of distinguished work by staff at Moray House".[32] Besides, the two-tier system of readerships and professorships referred to by Heriot-Watt could be avoided if the University agreed to the procedures now being proposed by Moray House. It was considered most unfortunate that the University had judged these to be unworthy of recognition even before it had seen them. Not only was that stance irrational, but it created the impression that the University was out of sympathy with Moray House's aspirations. What Moray House proposed was a set of arrangements that would fully meet the standards of external scrutiny and rigour which the University believed it currently operated. The letter then went on to consider the deteriorating relationship between the two institutions. It stated:

> *We have felt that Moray House and its Principal have been progressively marginalised by the University; associated college representatives have been excluded from important committees of the University; the Principals themselves are excluded from the work of the Principal and Deans' Committee, even though we know that matters relating to the associated colleges are discussed there.*[33]

Perhaps the most striking illustration of the University's attitude was provided by the University's refusal to share the contents of the draft report of the HEQC on the grounds that it was confidential to the University. The letter ended on a personal note in which the Principal stated that he had no personal interest in the severance of the linkage with the University and would prefer to find a way in which disagreements could be resolved amicably. In essence, what was sought was "a renegotiation of the relationship with the University". The matter was felt sufficiently important for Moray House to alert senior officers of the Scottish Office and the Funding Council to the fact that the deterioration in relationships with Heriot-Watt might become a matter for public discussion. In addition, the correspondence was reported to the meeting of governors on 10 December 1995.

That meeting of the governors therefore had two potentially explosive matters to consider: the correspondence on promotion, and the draft ordinance on accreditation. It would be fair to say that the Board was not enamoured of the University's accreditation proposal and the Chairman was asked "to convey its very serious concern about the University's proposals". The letter from the Chairman contained the following:

> *The Board is concerned in the first place that the scope of discussion will be restricted to the draft ordinance and will not cover the many issues on which representations have been made to the University over the past eighteen months. Secondly, the Board considered that the draft ordinance itself, contrary to the submission made by the principals and vice-principals of the associated colleges in March of this year, will require Moray House to adopt practices that are alien to its philosophy of professional education. Thirdly, no timetable has been set for the intended discussions. It appears that our opportunities to address matters will be indefinitely deferred when what they require is urgent action.*[34]

The Chairman went on:

> *I cannot stress enough the strength of feeling held by the Board and the senior staff of the college. There is a view that Heriot-Watt are not*

> *particularly concerned about the relationship and certainly are not prepared to give any priority to solving the problem.*

The Principal of the University replied early in January 1996 acknowledging, with regret, "that a conflict appears to have arisen in the relationship between our institutions"[35] and agreeing to an early meeting. Following discussions between the Chairmen of the two governing bodies, a meeting was held in early January 1996 with the Chairmen and the two Principals and it was agreed there that discussions between Heriot-Watt and its associated colleges would proceed in parallel.

The Final Round of Discussions

It was reasonable to expect that these discussions would prove fruitful, notwithstanding the seriousness of the disagreement that appeared to have arisen between the two institutions and notwithstanding also the serious inadequacy of the accreditation paper. On the one hand, there was pressure on Moray House to ensure that the academic linkage which it had sought would prove to be successful. On the other hand, the University Court, at its September meeting, had stipulated a series of guidelines for the discussion which provided further ground for optimism. These guidelines required the University's negotiators to proceed in a "cautious and conciliatory" way, rather than one characterised by "autocracy".[36] They also recognised that "the desire expressed by the colleges in recent discussions for greater academic autonomy provided an apparent impetus towards a changed form of association. The draft ordinance represented *one* (author's italics) possible mode of response to these aspirations". Moreover, the draft ordinance had been agreed "in principle", specifically noting that it would form only the basis for further negotiations with the several colleges. It was accepted that the document "would be susceptible to modification in the light of discussion with the colleges and that there was no requirement for a uniform approach to be adopted towards all three". In other words, the clear impression given by the Court discussion was that the University had recognised the full seriousness of the report of the HEQC; it had acknowledged that "the over-riding consideration from the University's point of view was the requirement for a mechanism to control academic standards"; and it had developed a mechanism on a purely provisional basis and hoped that, through collegial discussion, agreement could be reached in such a way that the University's concern about quality control would be met while, at the same time, the aspirations of the colleges would be realised.[37]

Regrettably, the series of meetings initiated in the early months of 1996 with Moray House did not fulfil these expectations. These were managed in a confrontational manner with the intention of undertaking a line-by-line scrutiny of the draft ordinance. From Moray House's perspective, what was sought was an open-ended discussion of accreditation, an academic arrangement clearly open to several interpretations, and, in Moray House's view, the interpretation adopted by the draft ordinance was completely at variance with the standard notion of accreditation as a process whereby an institution, as a whole, is given responsibility for the approval of its courses, subject to periodic auditing by the body whose degrees are awarded.

There were three meetings, the last of them on 8 May 1996. At that meeting, the University indicated that "accredited status could not be conferred immediately".[38] The University needed time to become familiar with Moray House's internal procedures and to acquire confidence in their effectiveness. When it was urged that "Moray House would also wish to be given a sufficient degree of responsibility for its own quality assurance arrangements", the University representatives maintained that they had no remit from Court to pursue that line of discussion and the matter would have to be referred back to the Court to receive authority to pursue a different course from that described in the draft ordinance. That was interpreted by the Moray House representatives as a gesture of ill-will and, with the talks clearly reaching an impasse, the Secretary of the University identified the options as he saw them. These were: merger, establishment of a faculty of the University, accreditation/validation, status quo, and discontinuation of association.[39]

The inclusion of the last option was significant. It could have been mentioned, of course, as part of a comprehensive listing of the possibilities. However, given the experience of three extremely disappointing meetings, the reluctance of Heriot-Watt negotiators to approach the discussion in a collegial and non-authoritarian manner, as well as the other evidence of institutional incompatibility, the suggested reference to the discontinuation of the association was interpretable as a hint that in Heriot-Watt's view severance of the linkage was the preferred option. As they returned to Moray House with nothing to show for a difficult round of negotiations and facing a host of unresolved problems, the Principal and Vice-Principal of Moray House agreed with the apparent Heriot-Watt assessment: there was no future for Moray House at Riccarton.

Reasons for Breakdown in Institutional Relationships

Several explanations might be offered for the breakdown in the relationship between Moray House and Heriot-Watt University.

Firstly, there was a failure to establish a mode of governance that was acceptable to Moray House. The college was an autonomous institution and it wished to establish a form of association that allowed it to retain significant autonomy. That was the central issue of concern in the discussions leading up to the publication of *Proposals for Linkage*, and it was at the heart of the paper *Towards a More Integrated University*. It also underlay Moray House's request to assume responsibility for the appointment of readers and professors. In Moray House's view, accreditation would give it significant responsibility for academic standards, while recognising its accountability to the University. Heriot-Watt University refused to countenance that proposal. It drafted an ordinance on accreditation without apparently fully recognising the difference between accreditation and validation. Its response to the highly critical HEQC report was to expect the associated colleges to acquiesce in an even closer degree of supervision by the University and to accept a validation relationship with the University that involved even tighter control than had been agreed in *Proposals for Linkage*. Moreover, the draft ordinance appeared to devalue Moray House's own well established quality assurance arrangements and expected it to agree to procedures for which it had long outgrown the need.

There was a failure here to acknowledge the precise nature of academic accountability. Rather than trust Moray House, with its significant experience of operating robust quality assurance arrangements, to exercise responsibility for the academic standards of its courses, with regular auditing of these approaches by the University, Heriot-Watt insisted that each course should be validated by a University committee. The University's response to the HEQC charge that it had not established an appropriate pattern of academic accountability in its associated colleges was to intensify the degree of direct supervision of its course review arrangements. Indeed, even when it was represented that a move of that kind, far from nurturing Moray House's concern for academic standards, actually removed responsibility for academic standards from the college and located it with Heriot-Watt University, the University insisted that arrangements had to be in line with the draft ordinance and, only belatedly, was persuaded to accept that a proper accreditation agreement might be adopted, but only after an interval of two years.

There was also a failure to acknowledge the importance of the principle of subsidiarity in academic governance. The case for subsidiarity, for devolving certain functions to the associated colleges, was strong. The associated colleges were independent institutions with very significant experience of exercising responsibility for the quality of their work. In any case, the Senate of the University did not have the range of academic expertise to evaluate standards in the three associated colleges: for that it looked to approved specialists drawn from other institutions across the UK. There was no reason whatever why the associated colleges could not have operated arrangements of that kind, including arrangements for the appointment of readers and professors, provided that these were conducted in accordance with rules agreed by the Senate and subject to audit by the Senate. The intransigence of the University on this issue became deeply alienating.

That alienation was intensified by the absence of an agreed way of handling disagreements. It is inevitable that in undertakings of any complexity, such as those involving collaboration between institutions of higher education, disagreements on policy and practice will emerge, and an effective academic alliance should allow these to be openly discussed and time given to their resolution. On too many occasions, once an issue had been raised, Moray House had to tolerate long and inexplicable delays before any action was taken. For example, Moray House petitioned the University to have a special graduation ceremony at Moray House for Malaysian students who were scheduled to leave for home ahead of the summer "congregation" at which awards would be conferred. No explanation was given, despite repeated requests, for the University's inability to reach a decision before five or six weeks had elapsed. Then again, when representations were made about the inappropriateness of Christian prayer at graduation ceremonies, not least when a significant number of the graduands were Moslem, it was assumed that, if a discussion was arranged between the Principal and the University Chaplain, all would be well. When that meeting proved inconclusive, the existing arrangements were continued and the same indignities were repeated at subsequent congregations. The idea of a meeting as a context for exploring disagreements as the necessary prelude to generating consensus seemed strangely foreign. In consequence, decisions taken by a Court or some other body in the University were seen as impositional and the decision-making process itself as authoritarian rather than collegial.

The correspondence on promotion illustrates this difficulty. Moray House certainly felt it had a grievance. Moreover, the issue was calculated to demonstrate to governors and to other members of the Moray House community that Heriot-Watt was unresponsive to the claim that some staff were being treated badly and that it was out of sympathy with Moray House's professional mission. Even if it is accepted that no university could possibly agree to the Moray House proposal that it award Heriot-Watt readerships and chairs, the Heriot-Watt response was a letter, extremely minatory in tone, requiring an answer which would be submitted to a subsequent Court meeting. Rather than engage in the discussion of the problem, the University adopted a line of action that would simply legislate the problem out of existence.

Secondly, it has to be acknowledged that the academic collaboration signalled in *Proposals for Linkage* simply did not materialise. It is true that the University worked closely with Moray House colleagues on the BEd in Technology and made a modest teaching contribution to the new course in Sport Sciences, and, of course, Moray House fully met the expenses involved in these contributions. Beyond these, however, there was no collaborative work. The University appeared to give such low priority to the development of a combined degree in Computer Studies and Education that, despite substantial efforts by Moray House staff, it did not see the light of day. In addition, despite the existence on the Moray House campus of the Scottish Interactive Technology Centre, and at Heriot-Watt of the Institute for Computer-based Learning, there were no initiatives to secure collaborative work in an area in which there was a clear duplication of expertise. Nor was it possible to turn to the validation arrangements as a way of raising standards. As the HEQC audit report noted, no value-added was identified for Moray House courses from the institution's involvement in the procedures established by Heriot-Watt University. That was partly because the University's course approval procedures were, at that time, embryonic, whereas those at Moray House, thanks to a long association with the CNAA, were well developed.

There are perhaps two reasons for the failure of the linkage to foster collaborative work. Firstly, Heriot-Watt's academic base was relatively narrow and there was an absence of "fit" with most Moray House activities. Secondly, the geographical distance between the institutions made it difficult to generate those interactions that are essential if genuine collaboration is to take place. The absence of an effective constitutional relationship between the two institutions might have been tolerable if, despite that, numerous joint academic initiatives flourished. The dearth of such developments raised questions about the value of continuing to seek an appropriate constitutional relationship.

Finally, on any dispassionate analysis, there were too many difficulties to allow a positive and lasting relationship between Moray House and Heriot-Watt University to evolve. Looked at from Heriot-Watt's perspective, the associated colleges were ambivalent: they wished to be part of the University community, when it suited them, while constantly asserting their autonomy. Nor can it have been easy for the University to handle three principals, who did not always speak with one voice, and who could be forceful advocates on behalf of their own institutions.

From the point of view of the associated colleges, the University's attitude was also ambivalent. Despite the public rhetoric about valuing the relationship with the associated colleges, it was widely acknowledged that not all members of the Heriot-Watt community favoured the development of closer ties with the associated colleges. One colleague was reported to have overheard the assertion that "It took Heriot-Watt more than 100 years to become a university and it is not going to be dragged down by the colleges". Whether or not that was a widely held view, there is evidence that members of the University did not wish to see relationships becoming too congenial. The exclusion of college staff, without consultation or explanation, from key University committees, and the failure of the University to share timeously the HEQC report on collaborative arrangements were examples of what came to be known as the "Fortress Riccarton" mentality. When that was combined with a degree of authoritarianism it was predictable that personal and professional relationships would deteriorate badly. And they did, to the extent that the trust and goodwill upon which institutional collaboration depends simply evaporated.

References

1. *The Times Educational Supplement Scotland*, 15 September 1989.
2. R E Bell, "The Education Departments in the Scottish Universities", Chapter 8 of *Scottish Culture and Scottish Education*, edited by Hamish Paterson and Walter M Humes, John Donald, Edinburgh, 1983.
3. Scottish Education Department, *Rationalisation of Secondary Teacher Training in Colleges of Education in Scotland*, March 1982.
4. Response of the Scottish Universities' Council for Studies in Education to the STEAC Report.
5. *The Times Educational Supplement Scotland*, 16 September 1988.
6. "Recognition must not be confused with validation", *The Times Educational*

Supplement Scotland, No 1075, 12 June 1987, page 2.

7. Ibid.
8. Nigel Grant and Roderick Lyall, "What is the great belittler so steamed up about?", *The Times Educational Supplement Scotland*, 4 November 1988, page 2.
9. Walter M Humes, *The Leadership Class in Scottish Education*, John Donald Publishers Limited, Edinburgh, 1986, Chapter 7.
10. Principal's confidential report to Heriot-Watt University, December 1989.
11. *The Glasgow Herald*, 7 July 1989.
12. *The Times Higher Educational Supplement*, 14 July 1989.
13. *The Times Higher Educational Supplement*, 7 July 1989.
14. *The Times Educational Supplement Scotland*, 30 June 1989.
15. *The Times Educational Supplement Scotland*, 15 September 1989.
16. *The Times Educational Supplement Scotland*, 7 July 1989.
17. *The Glasgow Herald*, 29 June 1989.
18. Letter from Professor Kirk to Professor MacFarlane dated 15 May 1992.
19. Minute of Special Meeting of the Moray House Board of Governors, 3 November 1992.
20. Paper from the Principal, Special Meeting of the Board of Governors, 29 January 1993.
21. Minute of the Special Meeting of the Moray House Board of Governors, 29 January 1993.
22. Minute of meeting of Court, 16 November 1992.
23. Letter from Professor MacFarlane to Professor Kirk, 3 March 1993.
24. Letter from Professor Rowan to the University, February 1995.
25. Paper agreed by the three associated colleges which accompanied the letter from Professor Kirk to Professor MacFarlane dated 3 March 1995.
26. Ibid.
27. The phrase was used repeatedly by Professor Knops at the June meeting of the University Joint Council.
28. Quality Audit Report on Moray House Institute of Education, Higher Education Quality Council, Quality Assurance Group, Birmingham, July 1996.
29. Paper from the Principal, Board of Governors, 12 September 1995.
30. Letter from the Principal of Moray House to the Secretary of the University, 25 September 1995.
31. Letter from the Secretary of the University to the Principal of Moray House, 5 October 1995.
32. Letter from the Principal of Moray House to the Secretary of the University, 13 October 1995.
33. Ibid.
34. Letter from the Chairman of Moray House Board of Governors to the Principal of Heriot-Watt University dated 20 December 1995.
35. Letter from the Principal of the University to the Chairman of the Board of Governors, 5 January 1996.
36. Minute of the meeting of Heriot-Watt University Court held on 25 September 1995.
37. Ibid.
38. Meeting between representatives of Heriot-Watt University and Moray House, 8 May 1996.
39. Ibid.

3

Faculty of Education

Introduction

Moray House's translation from Institute of Education of Heriot-Watt University to Faculty of Education of the University of Edinburgh was an unexpected and, for some, puzzling development. The public announcement that discussions had been initiated to effect a change of that kind had an even more startling impact on the educational and the wider community than the disclosure in 1989 that Moray House and Heriot-Watt University were contemplating a form of academic linkage. Commentators were intrigued, not simply by the switch of institutional allegiance, but also by the apparent conversion of the college of education which, more than any other, was perceived as the champion of the monotechnic principle and the fiercest opponent of "mergermania". This chapter offers an analysis of the institutional and educational matrix in which the change of allegiance originated; it describes the process through which a new institutional relationship was forged; it portrays the vision which inspired the merger discussions; and it examines the institutional framework and development strategy through which that vision is to be realised.

Dissatisfaction with the Status Quo

Chapter 2 sought to trace Moray House's growing disenchantment with its standing as associated college of Heriot-Watt University and the progressive deterioration in institutional and personal relationships. It came to be realised that not only was the existing relationship unsatisfactory but also that the prospects of negotiating an institutional arrangement acceptable to Moray House were bleak. Moreover, the legalistic and authoritarian handling of the last round of negotiations, in the course of which the Secretary of the University made a point of emphasising Moray House's right to discontinue the association, suggested that perhaps Heriot-Watt had no genuine commitment to its Institute of Education after all. Finally, it became apparent to anyone determined to take a close enough look that the promise of the linkage document had not been realised, and that, if anything, Moray House's institutional development was being thwarted. All of these considerations pointed to the conclusion that a different institutional partner was required.

Other considerations pointed to the same conclusion. At the time of the formation of the linkage with Heriot-Watt University the possibility had been mooted of the relocation of Moray House to the Riccarton campus. That was seen as a radical way of solving a perennial problem at Moray House, which concerned the inadequacy and, in some cases almost total unsuitability, of its existing accommodation and estate. One of the early initiatives of the Scottish Higher Education Funding Council was to require institutions to develop an estates strategy, on the understanding that property that was surplus to requirements might be sold and the proceeds retained, provided that an institution could demonstrate to the satisfaction of the SHEFC that these would be devoted to the further development of the estate. Moray House had significant surplus accommodation. It also received a substantial sum from the new SHEFC to enhance its library provision. The Moray House estates strategy, duly submitted to SHEFC in December 1994, committed it to a series of major capital projects funded by the sums accruing from the disposal of its surplus accommodation. That sequence of disposals and refurbishment projects was instrumental in the significant enhancement of Moray House facilities for teaching, learning and research. In consequence, Moray House was no longer dependent on a move to Riccarton to solve the institutional embarrassment of run-down and inadequate buildings.

Merger Pressures

The irretrievable breakdown in relationships with Heriot-Watt did not by itself constitute grounds for seeking a merger. Several features of the context of higher education in the mid-90s exerted pressure on institutions, particularly smaller and more specialised institutions, to consider their position. The first and most obvious of these was the intensifying financial pressure. While all institutions of higher education have been required in recent years to operate within increasingly tight financial constraints, for the autonomous teacher education institutions the financial environment throughout the '90s became increasingly hostile. The formula-based funding model, especially when combined with a continuing reduction in student numbers, placed the monotechnic institutions in particular difficulties. In order to avoid them suffering too large a reduction in income in any one year, they were "safety-netted", much to the annoyance of other members of the higher education community, who felt they were being expected to fund smaller and more expensive institutions. In Moray House's case the cumulative reduction between 1995 and 1997 was just under 9%, excluding inflation. That drop in funding had to be considered alongside the substantial reduction in funds accruing from in-service activities, which resulted from the Scottish Office decision to discontinue specific grant to education authorities for in-service work and the consequential decision of authorities to delegate funds for in-service to individual schools.

Faced with such difficulties and with no immediate prospect of their alleviation, it was necessary to consider the impact of intensifying pressures on staff morale and to question whether it was right that a wholly disproportionate amount of staff effort should be devoted to protecting the institution's financial position. In these circumstances, institutions like Moray House were bound to ask themselves whether the price of institutional autonomy and independence was a price worth paying. Could it be right to stand in the way of incorporation within a larger institution through which economies of scale would be secured and the work of the institution protected?

Evidence of the pressure on the institution is clear from the Moray House Strategic Plans of 1995 and 1996. The 1995 documentation made reference to "highly adverse financial circumstances" and acknowledged that, since that required progressive reductions in staff, it would be dependent on "safety-net" funding for a three-year period. That source of financial support was judged sufficient to stave off merger.[1] The decision of the Funding Council in 1996 to withdraw safety-net financial support created a totally different and discouraging scenario. In its Strategic Plan for that year Moray House was required to demonstrate how reliance on safety-net funding would be progressively eliminated. The plan read:

> *The Institute's analysis of its position . . . is that it must seriously question whether, as a relatively small institution, it can itself continue to provide the full range of services necessary to support its academic activities. If the arguments relating to economies of scale are valid, and the evidence is not conclusive, it is reasonable to postulate that some services can be provided more economically than at present. Securing year-on-year efficiencies may not be sufficient.*[2]

The conclusion was inescapable: the progressive erosion of the staffing and resource base called for "a strategic change of direction".

A second factor which made mergers more rather than less likely was the radical change in the context of higher education in Scotland caused by the Further and Higher Education (Scotland) Bill of 1991. The repatriation of the Scottish universities and the up-grading of the larger central institutions created a structure for higher education based on the universities. That undoubtedly led to the removal of an important obstacle to integration for it meant that all institutions came under the authority of a single minister. At a very early stage of its work, and in anticipation of future developments, the Scottish Higher Education Funding Council (SHEFC) endorsed a report on the criteria for evaluating mergers in higher education in Scotland. It singled out colleges of education for special mention, maintaining that "the committee does not see any positive advantages in colleges of education remaining essentially monotechnic institutions".[3] In addition, the SHEFC began to allocate funds specifically to support collaboration between institutions, particularly when that had long-term strategic significance for higher education as a whole.

Not surprisingly, shortly after the establishment of SHEFC, several institutional mergers took place: Jordanhill College with the University of

Strathclyde; Craigie College with the University of Paisley; Queen's College and Glasgow Polytechnic combined to establish Glasgow Caledonian University; Duncan of Jordanstone College of Art merged with the University of Dundee; and the Scottish College of Textiles sought merger with Heriot-Watt University. That pressure was accentuated when the Convener of COSHEP (Committee of Scottish Higher Education Principals) himself, John Arbuthnott of the University of Strathclyde, made a public statement to the effect that "the problems are particularly acute in Scotland's smaller higher education institutions. With more cuts to come there was a danger that some of these colleges would have to sacrifice their autonomy and merge with a bigger institution."[4]

The Educational Case for Merger

Important as these pressures were in placing merger on the Moray House agenda, a third factor was perhaps even more crucial. That concerned the educational case for merger. It was acknowledged that, internationally, teacher education flourished within a university context; that teachers should be educated in universities along with those entering the other great professions; that, nearer home, the post-merger evaluations from the west of Scotland demonstrated the academic and other benefits flowing from the incorporation of teacher education institutions within larger and more varied academic communities, even when these were located some distance from the host university[5]; and that staff and students could benefit enormously from participation in a quality of institutional life that small institutions by themselves simply could not sustain.

These arguments questioned the value of being merely an associated college, for they strongly implied that the advantages of membership of the university community were to be realised only through full incorporation rather than through the semi-detached status Moray House had with Heriot-Watt University. Tom Bone of Jordanhill had been a constant defender of that position during the STEAC review and beyond.[6] Then, reflecting on the merger of Jordanhill College with the University of Strathclyde, Arbuthnott and Bone referred to "the obvious possibility that, as the years went by, Scottish teachers who had been trained in a college rather than a university might be regarded as inferior in other parts of the world".[7] These limitations to associated college status, and the ways in which it could block aspirations, came to be acknowledged at Moray House on the basis of direct institutional experience. The same point has been recognised in subsequent merger discussions. For example, the St Andrew's College/Glasgow University merger proposal expressed a sense of frustration at "the lack of scope for development within monotechnics" and acknowledged the desire of the college "to develop diversified activities to provide sustainable growth and development in its academic and professional work through collaborative relationships with the range of departments and centres in the University of Glasgow".[8]

The Enhancement of Quality

A further factor that had a significant influence on the decision to consider merger concerned the determination to seek further improvements in the quality of Moray House's work. The Higher Education Quality Council (HEQC) auditors encountered "a vibrant staff and a well managed institution",[9] and their extremely positive report congratulated the Institute on its quality assurance processes and its strong links with the professions. Moreover, successive reports of the General Teaching Council for Scotland (GTC) and of the Scottish Community Education Council (SCEC) were extremely commendatory. Finally, the Institute's research strategy, led by a series of professorial appointments, was developing vigorously, and that was subsequently recognised when the Institute received the highest RAE rating of any of the former Scottish colleges of education in 1996. However, these external judgements clearly indicated that further progress was required if Moray House was to excel. Arguably, indeed, the rating of Highly Satisfactory for initial teacher education was a disappointing one for an institution whose raison d'être was teaching, as was the rating of Satisfactory for its social work provision. For all its strengths, Moray House needed the impetus that could be provided by full participation in the more varied and intellectually stimulating environment of a university.

Of course, the central challenge for all higher education institutions in the 1990s had been to maintain and enhance quality at a time of severe financial stringency. Indeed, institutions, individually and collectively, maintained that the tight funding regime itself represented a serious threat to quality. These pressures were felt more acutely by smaller institutions. However, it came to be acknowledged that in the prevailing climate, while it might be possible to maintain quality, it would become increasingly difficult to achieve the enhancements of quality that were sought, without significant institutional development of a kind that merger with a university would stimulate. It was reasonable to assume that teacher education programmes would be strengthened, academically and in other ways, through collaboration between teacher educators and appropriate specialists in the universities. It was also appreciated that the new faculties of education that would emerge in what might be called the second round of merger discussions, perhaps to an even greater degree than the earlier mergers, would create even stronger concentrations of expertise because they would bring together researchers and other education specialists from the host university, as well as the college of education, in a way that should significantly strengthen the research profile and output of the new faculties.

All of these educational benefits were subsequently encapsulated in Sir Stewart Sutherland's independent analysis of teacher education and training which was conducted as part of the Dearing/Garrick report and in which he was to say:

> *On academic grounds, I believe that the proposals that I have put forward . . . point, inevitably, towards the provision of teacher education within a broader intellectual context than can be provided by a monotechnic . . . There are also strong arguments that the student experience in Scotland is likely to be enhanced through being educated in a broader HE context and, indeed, that the staff experience will be strengthened through contact with, or integration into, the research environment of a university.*[10]

If these educational, institutional and economic considerations pointed so emphatically in favour of merger, why did Moray House not contemplate merger many years previously? How is Moray House's apparent U-turn to be explained? Firstly, the financial pressures on institutions like Moray House were much more severe in 1996 than they were ten years earlier. Indeed, it was reasonable for the Institute to declare its opposition to merger as recently as 1995, for it appeared that SHEFC's safety-net funding arrangements would serve to protect the institution's financial viability. Secondly, in the mid-80s, or even later, it was highly improbable that universities would have responded positively to overtures for merger. It is true that in the STEAC debate Alwyn Williams, Principal of Glasgow University, acknowledged that there was no objection in principle to including teacher education within universities. However, he subsequently confided to Tom Bone of Jordanhill College that the University would be most unlikely to agree to a merger with the college, as subsequent events were to prove.[11]

By the mid-90s there was evidence of a change of attitude on the part of the universities, with four of them offering awards in the field of teacher education. The 1992 legislation had repatriated the Scottish universities and emphasised that they were integral features of Scottish education. While they were members of an international community of scholarship, they also had strong regional and local responsibilities. Merger with a predominantly teacher education institution was an important way of recognising such responsibilities and would enable the universities to play an even stronger role in the education of teachers and their continuing professional development and, through that, to contribute even more strongly to the revitalisation of the work of the schools. Moreover, under pressure to strengthen the quality of their teaching, the universities could see the benefits of incorporating staff with strong experience of and commitment to teaching. Indeed, the establishment of faculties of education within universities was a way of rationalising a range of education-related activities which the universities themselves had developed over the years. Finally, the universities' change of attitude was influenced by the significant institutional development that had taken place in the colleges of education since the mid-'80s. Thus, for example, by the mid-'90s at Moray House all students were embarked on degree or postgraduate programmes of high quality, and the college had credible standing as a centre of research and development. Unquestionably, the commitment to merger at Moray House was strengthened by the realisation that, in the face of daunting financial pressures and through their own efforts, staff had succeeded in achieving standards in teaching and research which were now acknowledged to justify full membership of the university community.

Why not merger with Heriot-Watt?

The answer to that question must be obvious. If it was judged that relationships between Moray House and Heriot-Watt had irretrievably broken down, it is simply inconceivable that a merger could ever have been a serious proposition. What the linkage demonstrated was that there was a clash of cultures and an incompatibility of management styles. Not even a change of leadership could hope to remedy matters for, by the end of 1995, there was a strongly held conviction at Moray House that Heriot-Watt had proved to be an inappropriate partner.

However, even if relationships had been cordial, it is doubtful if a merger with Heriot-Watt would have proved workable. While mergers can withstand the difficulties that arise when the partners are geographically dispersed – for example, at Paisley and, to a lesser extent, at Strathclyde – the full benefits of merger are most likely to be realised when the institutions are sufficiently close geographically to allow academic interaction and exchange to intensify. In the Moray House case that would have required a relocation to the Riccarton campus and, the evidence was clear, there would have been little support for such a relocation, especially once Moray House's accommodation difficulties started to be addressed in a planned and systematic way.

On the other hand, merger with the University of Edinburgh had considerable attractions. That institution had had a long historical association with Moray House; it was geographically contiguous; it was a university that was internationally acclaimed; and its academic strengths across the social science and other disciplines strongly indicated that it could provide an appropriate context in which Moray House's future development could be fostered. What was proposed, therefore, was a change from what was described as a "semi-detached" relationship with a university to full membership of a university community, one that, moreover, offered numerous and varied points of contact and mutual support and therefore the prospects of strong academic collaboration and enrichment.

Besides, it was known that a review was being conducted of the provision of educational studies at the University, involving the most appropriate institutional arrangements for the Department of Education and the Centre for Educational Sociology. It was reasonable to assume that, given the pressure to establish concentrations of academic expertise, the difficulties associated with the rationalisation in the University might be eased when considered in the light of closer collaboration with Moray House. Finally, the Principal of the University of Edinburgh was known to have a strong academic and professional interest in education and in teacher education. That suggested that the University of Edinburgh might respond positively to overtures about future collaboration.

A range of factors, therefore, combined to place merger with the University of Edinburgh on the Moray House agenda: the relationship with Heriot-Watt had badly foundered; financial pressures were intensifying in a way that threatened the future of the institution as an independent entity; and significant enhancements of teaching and research were thought to be realisable through full incorporation with the internationally acclaimed university situated less than a mile away and that University, concerned as it was to strengthen its links with the rest of the Scottish educational system, could see numerous advantages in collaboration with Moray House. Of course, that collaboration could have taken many forms. However, in the first meeting between the two principals it was made apparent, as had been anticipated, that the only mode of institutional association that the University was willing to consider was merger. It was because both institutions were prepared to contemplate merger that the discussions were able to proceed.

Naturally, the Board of Governors was kept informed of the Heriot-Watt difficulties and was fully aware of the financial and other pressures on the institution. The Board's formal position, however, was that ways should be found of resolving these difficulties rather than adopting another partner, although at the meeting of the Board on 31 May 1996 there were several surprising expressions of dissatisfaction at the current state of affairs and of the need perhaps to make alternative arrangements. It was possible in meetings of the Chairman's Committee in February and September 1996 to have more candid exchanges on the options facing the institution. These exchanges, together with the confidential discussions with the Chairman of the Board, Peter Stillwell, and with the Vice-Principal, Jim O'Brien, confirmed the desirability and necessity of approaching the University of Edinburgh.

Joint Statement by the Two Principals

Early in July 1996, the Principal of Moray House approached the Secretary to the University, Martin Lowe, and asked him on a confidential basis if the Principal of the University of Edinburgh would welcome discussions on future collaboration. Subsequent contact with the Principal's Office indicated that a meeting would be welcomed. The first meeting took place on 30 July 1996. At that meeting the Principal of Moray House recounted the difficulties that were being experienced in the relationship with Heriot-Watt University and expressed a wish to enter discussions with the University of Edinburgh on a long-term strategic alliance. The Principal of the University responded positively but indicated that he would need to take soundings with colleagues. In the light of these discussions he would be in touch about the possibility of a future meeting. In the meantime, it was stressed that it would be imperative for Moray House to clear its lines with Heriot-Watt University, since the

University of Edinburgh did not wish to be accused of "poaching". Happily, the discussions that took place in the University were positive and a number of subsequent meetings took place at which the general features of collaboration between the two institutions were agreed as a basis for future discussion. Since the discussions between the two Principals had proved fruitful, a meeting of the Chairman's Committee of the Board of Governors was arranged for Friday, 18 October 1996. The paper before that meeting referred to the familiar difficulties that had been encountered with Heriot-Watt University; it maintained that "the future viability of the institution as an independent autonomous institution was open to question"; it recommended that merger with another institution of higher education seemed appropriate; and it considered that the University of Edinburgh "as a world class research-led institution", which had a long association with Moray House, would be an appropriate partner. The Chairman's Committee offered their wholehearted support for the continuation of dialogue with the University of Edinburgh.

In due course, on 23 October 1996, the Principal of the University of Edinburgh and the Principal of Moray House issued the following joint statement:

> *The Principals believe that both institutions, with their distinctive complementary strengths, wish to preserve and build upon all that is good in Scottish education and believe that such an integration will enrich and reinforce the contributions of both, strengthening and enhancing the contribution that each makes to school and professional education in Scotland, and securing these benefits into the next century.*[12]

They had agreed to commend to their governing bodies the proposition that the two institutions should consider a formal merger on the basis of the following principles:

(a) *undertakings to existing students will be honoured such that they will be in a position at least equivalent to that had no merger taken place;*
(b) *staff at Moray House with professorial and other University titles will retain these on integration into the University, with full appropriate consultation with relevant staff associations;*
(c) *the designation "Moray House Institute of Education" will be retained;*
(d) *Moray House's plans for the development of the Cramond campus would be taken into account in estates planning for the future;*
(e) *activities would be formalised within the University structure, possibly "by creating an additional Faculty with appropriate planning and budgetary responsibilities";*
(f) *both institutions would approach merger with a staffing profile appropriate to their activities and recurrent budgets at point of merger;*

(g) *each institution commits itself to full disclosure with regard to current and projected financial position;*
(h) *interim arrangements will be agreed with Heriot-Watt University.*

The statement continued:

> *The fundamental objective of the proposed merger is to maximise the academic benefits bringing together the activities of Moray House Institute with those of the University of Edinburgh. The merger is being undertaken in accord with the SHEFC principles of securing greater efficiency and effectiveness in the higher education system, while safeguarding and enhancing quality of teaching and research.*

And concluded:

> *We both believe that there are good educational and structural reasons for effecting this proposal, but we are also concerned to see this fully debated by the governing bodies and the community of each institution. For this reason, we have chosen to make this statement public at an early stage as a personal declaration of a view we both share. Naturally, we hope these discussions will move to a positive conclusion but are not in a position to pre-commit our governing bodies.*

The Public Reaction

The joint announcement attracted significant media attention. *The Herald* headlined "Merger plan for giant campus". According to *The Scotsman*, "The merger would also help Edinburgh secure a place in the proposed 'super league' of universities". *The Dundee Courier* anticipated the establishment of "a super university"; *The Press and Journal* made reference to the creation of "one of the UK's largest universities"; *The Evening News* highlighted the "economies and efficiencies that will develop from the integration" and the way in which Moray House will flourish "more broadly within the University of Edinburgh."

The Times Educational Supplement Scotland, having quoted the Principal of Moray House as claiming that "Survival is not a laudable aspiration. I am paid to make my institution flourish", offered the following editorial comment:

> *A merger between Edinburgh's two oldest academic institutions has made sense for many years, but the proposed announcement this week by the principals of Edinburgh University and Moray House Institute owes as much to financial pragmatism as to scholarly opportunity . . . Neither of the two Edinburgh institutions is likely to reject their principals' lead. The university stands to gain by increasing its under-*

> *graduate numbers, by complementing its own research strength (symbolised in the new Institute for the Study of Education and Society) with the school-oriented work of Moray House, and by contributing to teachers' professional development, especially through higher degrees. Moray House, having increased its own research efforts, will benefit not just from the university's education staff but from those in other disciplines. It will also gain access to and credibility from a university with international standing . . . Academically, it makes more sense for Moray House to go in with Edinburgh.*[13]

The same journal resumed its consideration of the matter in a subsequent issue. It considered that "Moray House shows the way" and saw significant advantages for staff in working "in Scotland's most successful university". The hope was expressed that the financially driven constraints of an institution largely dependent on one declining activity would be removed and that Moray House would approach the merger "not as a client seeking help but as a junior partner".[14]

The Heriot-Watt Reaction

While the announcement of the merger discussions received enormous support both inside Moray House and outside the institution, it was to be expected that it would not be received with acclamation at Heriot-Watt University. The press release from that university attempted to be neutral in tone but was clearly an expression of disappointment. It read:

> *We have only recently become aware of the approach by Moray House to the University of Edinburgh and regret there were no three-way discussions to ensure that, whatever final decisions were reached, they would be in the best interests of all three institutions, teacher training in Scotland, and of all our students, now and in the years to come. . . We have willingly worked with them as an associated college, contributing to the significant raising of standards and collaborating on many new degree courses.*[15]

The more widely quoted responses in the press referred to "shock" and "surprise". The University claimed that Moray House's decision to enter discussions with the University of Edinburgh "flew in the face of planned developments",[16] a reference to the distinctly unpromising discussions on accreditation held earlier in the year. The University also maintained that it was "a gross discourtesy"[17] to be kept in the dark about these discussions. Indeed, with the possible intention of discrediting Moray House and its Principal, concern was expressed at the lack of consultation prior to the announcement and at the clandestine nature of the Moray House move, with the implication of duplicity in engaging in discussions with Heriot-Watt on the one hand, while simultaneously pursuing separate discussions with another potential partner. At the press conference on 23 October, reference had been made to the possibility of enabling Moray House students to "access scholars of international distinction" at the University of Edinburgh. Taken out of context, that was read as a "calculated insult" as far as Heriot-Watt was concerned.[18] To make matters worse, Heriot-Watt was repeatedly referred to in the press as being "snubbed", "rejected", "left out in the cold", instead of being regarded as a legitimate participant in "three-way discussions".

It is understandable that Heriot-Watt should present itself as an aggrieved partner, reacting to the Moray House announcement "more in sorrow than in anger", and that a degree of sympathy should be expressed on its behalf, as, for example, in *The Times Educational Supplement Scotland*, which, while acknowledging that it was right that Moray House should merge with Edinburgh, felt that "Heriot-Watt can justifiably feel sore".[19] Setting these feelings of disappointment to one side, there are several responses to the claims made in the Heriot-Watt reaction.

Firstly, the sense of shock and surprise that was claimed was disingenuous. Correspondence between the two institutions over the previous eighteen months, as well as exchanges at meetings, clearly indicated that relationships were deteriorating badly. There were references in that correspondence to "the problems that exist between our two institutions"[20]; there were claims that "some very real difficulties will have to be overcome"[21]; mention was made of the "quite vindictive tactics that were being deployed by the University" and to difficulties which, despite the passage of time, remained "undiscussed and unresolved".[22] The signs were clearly there of Moray House's growing exasperation and it is simply inconceivable that Heriot-Watt was not aware of quite widespread disaffection.

Secondly, there was no duplicity. The decision to approach the University of Edinburgh was made in early July 1996 and the first meeting was scheduled for the end of that month. It is true that in the intervening days the Principal Designate of Heriot-Watt University visited Moray House, and was alerted to the existence of serious difficulties. It is naïve to expect that he should have been informed during such an introductory meeting that discussions were about to open with the University of Edinburgh. Common prudence suggested that the time to approach Heriot-Watt University was after it had been established that a positive response to merger discussions had been received from the University of Edinburgh. It was crucially important not to give Heriot-Watt University any opportunity to sabotage talks before they had commenced.

Thirdly, it may be legitimate to claim that Heriot-Watt contributed to the raising of standards at Moray House, but it is also legitimate to question the extent and nature of that contribution. As the HEQC report demonstrated, Heriot-Watt's influence on quality assurance arrangements at Moray House was tenuous at best, not least because its own quality assurance arrangements were relatively under-developed. In the drive to raise standards, Heriot-Watt was almost entirely dependent on specialists drawn from other institutions, as members of the Academic Advisory Committee. It was always Moray House's claim that it could have managed that process itself, and probably a good deal more efficiently. Besides, the key to quality enhancement is academic interaction and exchange and that only occurred to any significant extent in the BEd in Technology, and to a much lesser extent in the BSc in Sport Sciences. The charge that Heriot-Watt was badly treated would only be valid if the University's contribution to the enhancement of Moray House's activities was significant and substantial: it was nothing of the sort.

Finally, the Moray House Board made it emphatically clear that Heriot-Watt University had no entitlement to participate in "three-way discussions". A letter to the Principal of Heriot-Watt University stated:

> *We are an independent and autonomous institution: responsibility for making decisions about our future rests squarely with us and, in discharging that responsibility, we have no accountability to Heriot-Watt University.*[23]

The same message was communicated in a letter to the Scottish Higher Education Funding Council and to the Scottish Office, affirming that "the Moray House Board of Governors . . . cannot concede that Heriot-Watt University has any locus in discussions about our strategic development".[24]

The Consultation Process

Following the joint statement by the Principals of Moray House and the University of Edinburgh there was an extensive consultation exercise in both institutions, and a special meeting of the Board of Governors was arranged for 29 November 1996 at which the Board received a report. That report intimated that at a special meeting on Friday, 22 November 1996, the Academic Board had "unanimously agreed to recommend to the Board of Governors that formal discussion should take place with the University of Edinburgh". It also referred to unsolicited letters from heads of educational institutions, from colleagues in other teacher education centres, from the Director of Education of the City of Edinburgh, and the Registrar of the General Teaching Council. It concluded: "The overwhelming response from the consultation is strongly positive."

The report to the Board of Governors justified the merger discussions with the University in these terms:

> *Moray House had a long historical association with the University of Edinburgh. Its academic scope provides an appropriate context for the professional programmes for which Moray House is well known. The University of Edinburgh has scholars of the highest distinction in fields*

directly relevant to Moray House's work. In addition, the University of Edinburgh is a research-led institution of the highest distinction. These considerations suggest that the future well-being of Moray House could be secured through full incorporation in that University. Through that incorporation the quality of our teaching will be enhanced and our research work significantly strengthened.[25]

The Board unanimously agreed that discussions should begin and, within a week, a similar decision was taken by the appropriate bodies in the University of Edinburgh. Subsequently, extensive discussions took place between staff from different parts of both institutions and a number of "due diligence" studies were undertaken. In due course, a "Joint Liaison Group" was established which, since it consisted of four senior colleagues from each institution, became known as the "4 + 4 Group". Its purpose was to oversee the merger discussions and to constitute a means of reporting to both institutions.

The discussions proceeded sufficiently well for a report to be submitted to the March meeting of Governors. It had two parts: the first was an integration paper which set out the proposed constitutional framework. That saw Moray House being established as a Faculty of Education which would incorporate education specialists in the University; the discontinuation of the Moray House Board of Governors; the possible relocation of activities at the Cramond campus; and the establishment of an agreed management structure for the new Faculty. The central thrust of the integration document emphasised the academic advantages of merger:

The proposal is made on academic grounds and envisages that Moray House Institute will be fully and transparently integrated into the University of Edinburgh which will, in its enlarged form, include as an integral part an Institute of Education focused on excellence in both research and professional practice, and having regard to the needs and development of education in Scotland.[26]

The same meeting of the Board of Governors approved a Memorandum of Understanding, which set out a number of general principles concerning the merger discussions as well as the membership of five joint working groups that would undertake detailed studies of academic matters, estates, finance, services, and staffing. There was a commitment to generating a single merger document to be submitted to the Secretary of State by late summer, with a view to merger from 1 August 1998.

The discussion at the Board of Governors on 18 March 1997 was interrupted by news of a press release issued by the Principal of Heriot-Watt University in which he made a further plea for the involvement of Heriot-Watt University in three-way discussions on the future of Moray House. That press release reported an address the Principal had given to the

General Council of Heriot-Watt University which, in turn, was based on a letter he had sent to the Principal of the University of Edinburgh earlier in March 1997. The central thrust of the letter was to claim that Heriot-Watt University had a strong involvement in a number of Moray House courses; that it had made significant academic investments in the association with Moray House; and that it had a particularly close relationship with staff and students at the Cramond campus. The letter culminated in a claim that it would be appropriate if the University of Edinburgh were to incorporate the Holyrood campus, while the Cramond campus might be incorporated by Heriot-Watt University. That, it was concluded, would represent a "satisfactory split" of Moray House.[27] The Board of Governors was infuriated by that press release and its appearance simply served to reinforce the resolute commitment to insist that Moray House, in its entirety, would merge with the University of Edinburgh, and that Heriot-Watt would have no part in these discussions.

That interjection by the Principal of Heriot-Watt was interpreted as a clear attempt to destabilise the discussions between Moray House and the University of Edinburgh. Moreover, the claims made about investment in Moray House and the development of the Cramond campus were spurious. Indeed, it would be fair to say that in the course of the association with Heriot-Watt there had been no discussion of a strategic kind on academic developments. It is true that Heriot-Watt University had an involvement in the development and teaching of programmes based at the Cramond campus. However, the teaching commitment totalled precisely 125 hours in the BSc in Applied Sport Sciences. It had no part in the BEd in Physical Education, in the BA in Leisure Studies, or in any of the eight master's degree programmes based at Cramond. Heriot-Watt University had certainly been involved in the teaching of the BEd in Technology. However, there was a transfer of funds from Moray House to Heriot-Watt University to ensure that staff were appointed to support that work. In permitting that degree of collaboration, Moray House was complying with the national guidelines which stipulated that a university institution should provide the technology input to that degree.

It was not difficult to see in the correspondence from Professor Archer a clear determination not to act in the best interests of Moray House but to protect those of Heriot-Watt University. Indeed, the letter to the University of Edinburgh concluded with the hope that "it may still be possible to salvage something honourable from the present situation". The Moray House reply was indignant:

> *Perhaps the most disappointing feature of your letter is the attitude it portrays towards Moray House. Our interests and concerns appear to be of little consequence. You imply in your revealing metaphor of "salvaging something" that Moray House is simply there to be exploited in furtherance of the interests of Heriot-Watt University. Given that attitude, I am sure you will fully understand why we*

> *cannot agree to Heriot-Watt University's involvement in discussions about our future.*[28]

The references to "salvaging" and to the achievement of "a satisfactory split" of Moray House caused very considerable offence, and the Principal of Moray House made a public statement to the effect that the Heriot-Watt interjection had to be interpreted as a "hostile act". He maintained that, in its relationship with Heriot-Watt, Moray House "had run up against a series of grievances which Heriot-Watt proved incapable of resolving, so we have no intention now of engaging in tripartite talks".[29]

Moray House's angry response to these interjections received wide publicity. Representing the views of the Board of Governors, the Chairman affirmed that the Board was committed to merger with the University of Edinburgh, meaning by that "the full incorporation of Moray House, in its entirety, within the University". His letter went on:

> *I am pleased to say that steady progress is being made in the merger discussions and both institutions have agreed how that integration will be effected so that all our courses, including those based at Cramond, will become the responsibility of the University of Edinburgh. Consequently, there is no question of a "satisfactory split" of Moray House and "tripartite working parties" have no function to perform. In any event, my Board could never agree to the dismemberment of the institution and cannot concede that Heriot-Watt University has any locus in discussions about our strategic development.*[30]

The press release issued immediately following the March meeting of the Board of Governors took the same consistent line. It stated that

> *. . . because its experience with Heriot-Watt University had been unsatisfactory in many respects, it had decided to seek a new partner. At their meeting today the Board of Governors confirmed their original decision that Heriot-Watt would have no part to play in the merger discussions.*[31]

Fortunately, on this matter, the University of Edinburgh and Moray House stood in strong agreement. For example, on 17 March 1997 the Secretary to the University wrote to Professor Archer in the following terms:

> *The University of Edinburgh is simply not in a position to address separate proposals for the future of the Scottish Centre (at Cramond). The basis on which the Principal of the Institute sought merger talks, and which the University has consistently honoured, was that these would involve the whole Institute, it being the firm commitment of*

> *the Institute's governors to retain the Centre as an integral part of its provision.*

At the same time, the following comprehensive statement was provided by the University:

> *The present discussions between the University and Moray House Institute of Education were instituted at the request of the Principal of Moray House, who had taken the view that the Institute's future activities would best be pursued via a full merger with the University of Edinburgh, with the range of disciplines and facilities it offered, rather than through the somewhat looser validating and associated arrangements that had applied for the past few years with Heriot-Watt University. Since then, the project has been welcomed by the governing bodies of both the University of Edinburgh and Moray House Institute of Education, and a framework for progressing the merger and a Memorandum of Understanding have this week again been endorsed by our governing bodies.*[32]

Not deflected by the statements from Heriot-Watt University, both institutions committed themselves to more detailed merger discussions. The 4 + 4 Group received in turn reports from the sub-groups that had been created to deal with staffing, estates, finance, and administrative/academic/student/ support services. In due course, a draft report was submitted to a special meeting of the Moray House Board of Governors on 10 July 1997, along with the response to a paper which had been prepared for internal consultation at Moray House. The Board of Governors unanimously agreed the terms of the merger document and arrangements were made for a submission to be made to the Secretary of State in August/September 1997, proposing a merger date of 1 August 1998.

A Proposal for Merger

The merger document set out a vision of an enlarged institution which would include:

> *. . . as an integral part an institute of education focused on excellence in both research and professional practice and having particular regard to the needs and development of education in Scotland.*[33]

It was acknowledged, of course, that the merger offered reciprocal benefits. On the one hand, it was claimed, the merger "will substantially enrich the education of teachers and members of kindred professions through the strengthening of the academic underpinning of their education, giving them a broader cognitive base from which to draw in their work".[34] By the same token, the merger would enable the enlarged University "to have a significant impact on the professional education of teachers" and "to play its part in the revitalisation of the work of schools and other educational agencies". Through the merger it was concluded that the University would be able "to strengthen and, at the same time, to be invigorated by, its contact with schools and other agencies".

Given the fundamental objective of maximising academic benefits by bringing together the activities of both institutions, the merger document set the following strategic objectives:

1. To establish a major high quality institute of education for Scotland thereby raising the quality of teaching and research in Education.
2. To strengthen and broaden the academic base of students in Education and other related fields within the merged institution.
3. To strengthen and broaden research in Education and other cognate areas in the merged institution.
4. To provide a framework for the development of continuing and professional education for staff in education at all levels, from pre-school to higher education.
5. to maximise opportunities for all staff and students.
6. To widen access to undergraduate programmes, particularly for Scottish students.
7. to enhance contributions to school education.
8. To provide a more efficient and effective support service thereby freeing resources for investment in the academic base.

How was that academic vision to be realised in institutional terms? The merger settlement entailed the full integration of Moray House Institute of Education within the University of Edinburgh. The Moray House Board would be disbanded, with all its responsibilities and liabilities transferred to the Court of the University, which would be the single employer of all staff, and which would include representation from the Moray House Board. A new Faculty of Education would be established with full planning and budgetary responsibilities for the first four years of its life, before joining the Faculty Group currently consisting of the Faculties of Law and Social Sciences. The merger was based on the integrationist principle that similar units should be grouped within a single management framework: the Moray House Social Work Department would therefore join the University's Social Work Department; the two units that provided programmes in the teaching of English to speakers of other languages would move forward in a framework of collaboration; and the various administrative and support services would be integrated to form, for example, a single registry, a single finance department, a single building and estates department, and so on, with each of these providing an integrated service operating according to University of Edinburgh standards and managed from a single location on the basis of rationalisation of provision rather than its centralisation. Finally, the new Faculty would involve staff from Moray House as well as established specialists in the University's Institute for the Study of Education and Society and the Centre for Teaching, Learning and Assessment in Higher Education and through that integration would create a strong concentration of expertise in the academic and professional study of education.

The Moray House title would be retained "in recognition of its international reputation and its familiarity to schools and the wider educational community in Scotland". The existing Principal of Moray House would be Dean of the new Faculty with the responsibility of a Provost and membership of the University's Central Management Group. The other senior academic posts would be filled by the existing Vice-Principal, who would be Vice-Dean; the existing Senior Assistant Principal, who would be Associate Dean; a Professor of Education, who would be Associate Dean with responsibility for research policy; and another Professor would carry responsibility for postgraduate studies.

The new Faculty would replicate the committee structure of the other faculties at the University of Edinburgh. It would consist of eight academic departments, and a commitment was given to establish a Graduate School, similar to that found in other international centres of excellence in education, to serve as a centre for postgraduate study and research. Finally, again in line with practice elsewhere in the University, there would be established a Faculty Office to provide essential administrative support to the new Faculty. Organised in this way, the new Faculty was charged to pursue excellence in teaching and research; to retain its strong links with the professions and with professional practice; and to continue to provide support of various forms for the educational service.

A further example of the integrationist philosophy underpinning the merger concerned the relocation of activities based at the Cramond campus to the city centre, subject to independent advice on feasibility and cost. Accordingly, an independent assessment, financed from the SHEFC Strategic Change Fund, was commissioned. That identified five options as follows:

1. status quo;
2. integration into the University of Edinburgh and disposal of both Holyrood and Cramond campuses;
3. retention of Holyrood and disposal of Cramond;
4. retention of Cramond and disposal of Holyrood;
5. retain both Holyrood and Cramond.

Option 2 was not pursued, since the evidence clearly indicated that there was no surplus accommodation at the University of Edinburgh. The analysis of the remaining options was conclusive: it demonstrated that, from a financial point of view, option 3 was overwhelmingly superior. The attractiveness of that option was supported by other considerations. Firstly, it was accepted that a city centre location would be attractive to very many students. Secondly, the existence within the University of Edinburgh and at the Holyrood campus of Moray House of substantial facilities to support physical education and related studies strengthened the case for relocation. Thirdly, it was believed that the educational case for integration would be strengthened by physical integration. The case for the merger rested substantially on educational grounds. It was maintained that there were reciprocal benefits to be derived from the collaborative activity which the merger would make possible; that research at Moray House would be enriched through the creation of a critical mass of educational researchers and through providing Moray House researchers with access to the infrastructure of a research-led University; that Moray House courses would be strengthened through the involvement of scholars from a wide range of disciplines; and that staff and students at Moray House would benefit from participation in the life and work of an internationally acclaimed University. These arguments were also seen to apply to the Scottish Centre for Physical Education, Movement and Leisure Studies. There was a distinct possibility that, if it were retained on its existing site, it would lose many of the benefits which the merger sought to realise. Accordingly, both institutions committed themselves to the full incorporation of the Scottish Centre within the University and to its further development as a centre of national and international standing. The attractiveness of the relocation of Cramond-based activities was reinforced when it became apparent that the Moray House bid for a major development on the Cramond campus would not be supported.

The relocation option was based on the assumption that the existing facilities at Holyrood, the Pleasance and Peffermill would not, as they stood, be sufficient to provide specialist facilities for physical educationists and students in related disciplines. It would be essential to ensure that, prior to the relocation, appropriate new and enhanced facilities would be made available in the city centre. As the merger discussions proceeded, an ambitious plan for estates development was agreed, involving enhancement of the playing field accommodation at Peffermill, a new swimming pool at Holyrood, the enhancement of the existing physical education and sports complexes at Holyrood and the Pleasance, the extension of the Holyrood library, and the establishment of a new block to support academic and teaching activities. These developments would be financed through funds accruing from the disposal of Cramond and other assets, with the University taking the opportunity to upgrade some of its recreational facilities in the Pleasance area as part of a comprehensive development scheme.

The Period of Implementation

One of the features of the merger settlement was that there would be a four-year period of implementation during which there would be full harmonisation of staffing, academic and administrative procedures. Accordingly, three milestones were established, with targets set for each. The first of these was the merger date itself. By 1 August 1998 the Moray House Board would have been wound up; the Faculty of Education would be created; the senior management of the new Faculty would be in place, along with the full committee structure; the Social Work undergraduate degree programme would be transferred; and the new departmental structure established. Finally, the Faculty Office and full support system would be in operation. For the period to 1 August 1999, that is, one year after the merger, Moray House's undergraduate programmes would be revised to accord with provision in other faculties of the University of Edinburgh; a Graduate School would be established; reasonable progress would be made on the teaching strategy and on the research strategy; and a plan would be agreed for the relocation of Cramond-based activities. Finally, by 1 August 2002, the end of the transition period, full integration would be established with the new Faculty operating in line with practice in other faculties; and the relocation of the Cramond campus would be completed. To oversee the implementation strategy a Merger Implementation Group was established, chaired by a Vice-Principal and reporting regularly to the Central Management Group.

Academic Strategy

The academic strategy following the merger covered both teaching and research. With regard to the first of these, there was a commitment to ensuring that both undergraduate and postgraduate provision would be strengthened. The following were the main features of the teaching strategy:

(a) Restructuring of undergraduate provision to bring it into line with provision elsewhere in the University of Edinburgh.
(b) Transfer of social work provision to the Faculty of Social Sciences.
(c) A framework for collaboration between Moray House's provision in English Language Teaching with similar provision at the University of Edinburgh.
(d) In the postgraduate domain, there was a commitment to the rationalisation of Moray House's wide-ranging Modular Master's Scheme with a view to the elimination of duplication and to the provision of support for continuing professional development on a more cost-effective basis.
(e) The development of masters' degree provision would be significantly extended by the establishment of a Graduate School, which would, in due course, achieve an international reputation and attract postgraduate students and others. Moreover, postgraduate links would be established with other faculties, with the intention of significantly enhancing the number of postgraduate students.

The strategy for research, since it would bring together within the one entity established members of staff at the University together with Moray House staff, was to create an enhanced research facility with the aim of securing a 4 rating in both units of assessment in the next Research Assessment Exercise scheduled for 2001. A commitment was given to secure the greater involvement of staff in research activity, with a more substantial output of published work, focusing on a restricted number of areas in order to cultivate genuine research strengths. In addition, it was planned to strengthen the research infrastructure in the new faculty by increasing the number of research assistants and other research staff, including contract research staff, whose contribution to the creation of a lively research culture was fully recognised.

Of course, it was recognised that there can be a tension between teaching and research. In line with practice in other colleges of education, the extent of staff involvement in research at Moray House had been limited, although it was developing. The merger committed the new Faculty to a significant enhancement of its research profile without, in any way, questioning the commitment to teaching or the commitment to enhancing the educational experience of students.

Evaluation of the Merger Proposal

There are several tests of the validity of a merger proposal. The Shaw Report of 1991[35] identified four criteria which were subsequently accepted by the Secretary of State. These were

(a) achieving wider access to and within higher education;
(b) education provision must be enriched and widened;
(c) safeguarding and improving quality, relevance and efficiency;
(d) retaining local initiatives and accountability in meeting local needs whilst maintaining and reinforcing national and international competitiveness.

With regard to the first of these, it was maintained that the merger might have a significant impact on recruitment of males to the Primary BEd but would certainly extend educational opportunities to students recruited to Moray House, chiefly because of the more flexible course provision that is available within an enlarged institution. With regard to the second criterion, it was maintained that the merger would lead to a significant enrichment of course provision through the involvement of the University staff in teaching and through the creation of collaborative programmes that would not have been possible if the institutions continued to act separately. The third criterion was met by maintaining that both teaching and research would be enriched because these would be conducted within a research-led University, while significant economies would be achieved through the integration of services, through economies of teaching, and through the integration of estates. Finally, with regard to the fourth criterion, it was maintained that the University of Edinburgh, while being a member of an international community of scholarship, was located in a particular community. The merger with Moray House was seen as a way of embedding the University even more firmly in that community, of extending the University's links with the professional communities, education authorities, and other agencies in the east of Scotland and beyond, and of enabling it to play a more influential role in the revitalisation of the work of the schools and the provision of support for the professions. The creation of a single institution with a strong education faculty was thought to establish a single door access in the east of Scotland to a very considerable community of expertise in professional education and research.

Of course, it was necessary to ensure that appropriate transitional arrangements for students were agreed and it was encouraging that Heriot-Watt University responded positively and generously on this matter. In the end, agreement was reached that all students entering the University from October 1998 would be registered students of the University of Edinburgh and, on completion of their studies, would receive awards from that university; that students in their final year of study in 1998 would obtain awards of Heriot-Watt University; and that students then in their first, second or third year of study would have the option of graduating with a degree of Heriot-Watt University or the University of Edinburgh. Subsequent discussions have confirmed the validity of the decision to engage in merger discussions with the University of Edinburgh since overwhelmingly students have expressed a preference for graduation from the University of Edinburgh.

The collaboration secured with Heriot-Watt University with regard to the transfer of students was reinforced in the good wishes expressed by the Chancellor of Heriot-Watt University, Lord MacKay, in his graduation address on 9 July 1998. At that ceremony, Lord MacKay paid handsome tribute to the work of Moray House and wished it well in its relationship with its new partner. The same generous expression of support was provided by Professor Charles Brown at the final Moray House graduation ceremony in November 1998. Finally, the merger discussions did not rule out the possibility of future collaboration with Heriot-Watt. The University of Edinburgh already has a number of programmes developed jointly with Heriot-Watt University, and there is an openness to consider ways in which collaboration might take place in areas where it is agreed that Heriot-Watt University has an important contribution to make.

Epilogue and Prologue

The merger proposal was agreed by both institutions in September 1997 and lodged with the Secretary of State then, with an anticipated merger date of 1 August 1998. Eventually, in June 1998, the Secretary of State formally agreed the merger of the two institutions. It had been a huge undertaking and was concluded according to the agreed deadline. Even before the end of 1996, discussions had been held with the Mergers Committee of the SHEFC, which served to convince the institutions that there appeared to be no major obstacle to the merger. They therefore decided to proceed on the assumption that the merger would, indeed, take place. From the date of the submission of the merger proposal and extending into 1998, detailed plans were taken forward in line with the implementation strategy.

It was clear to all, then, that 1998 would be a landmark in the history of Moray House: it was the year in which it would be formally integrated into the University of Edinburgh. However, 1998 was significant in another sense: it marked 150 years of professional education at the Holyrood site, the former Edinburgh Training Centre having moved there in 1848. In anticipation of that landmark, plans had been made for the 150th Anniversary Programme of activities that would extend throughout the first half of 1998. That consisted of music and dance, some of it specially commissioned from staff and students, and drama; it involved the establishment of a remarkable archive of the institution; and, through the 150th Anniversary Appeal, it facilitated the enhancement of the campus environment, which was highlighted by a specially crafted centrepiece, a sundial bearing the Moray House crest, *Nosce Te*, to remind future generations of students of the perennial injunction to *Know Thyself*; and it reached a climax in a magnificent procession down Edinburgh's High Street, with present and former staff and students of the college and its Demonstration School re-enacting the journey taken by their counterparts in 1848, to be followed by a service in the Canongate Kirk, in which tribute was paid to a great tradition of professional education. These various forms of celebration had several purposes: they permitted past and present members of the Moray House community to record a landmark in the history of the institution; they contributed to Scotland's tradition of music and dance; they bequeathed an enduring legacy of Moray House's achievements; and they were a means by which staff asserted their professional and institutional allegiance.

The 150th Anniversary sought to celebrate Moray House's contribution, and that of two other colleges, Callendar Park College of Education and Dunfermline College of Physical Education, to the educational life of Scotland. It was also a fitting way of drawing one chapter in the history of the institution to a close as a prelude to the significant change that the merger would entail. The celebrations assumed an added poignancy from the realisation that the old order was giving way to new. It was, however, fully recognised that, though the institutional context would change, the Moray House tradition, with its distinctive professional mission, would be preserved.

The University of Edinburgh, itself a strong and committed ally of professional education, recognised the need for that kind of continuity. It agreed to the retention of the Moray House title, not as an assertion of institutional independence, or as evidence of reluctant participation in the enlarged University, but as a continuing reminder of what Moray House stood for. That theme was invoked in the ceremony in which the merger was formally celebrated on 29 September 1998. The address from the Moray House Principal concluded as follows:

> *All of us who are formally received into the University community today are the inheritors and therefore the custodians of a distinguished tradition of professional education. Responsibility for nurturing that tradition now rests with the University of Edinburgh. However, professional education is such a strongly embedded feature of the work of the University that it can readily assume that extended responsibility. We therefore honour our obligation to the Moray House tradition by committing ourselves to the service of the University of Edinburgh. On behalf of all my Moray House colleagues I gladly declare that commitment today. We pledge ourselves to work for our full integration into the University; we express our allegiance to the values that are at the heart of the University; we identify with the academic aspirations of the University and the codes of academic practice which sustain it; and we commit ourselves in all our activities to make our distinctive contribution to the development and enhanced standing of the University. And we give these undertakings with confidence and pride.*

With these words, in the majestic setting of the University's Playfair Library Hall, Moray House was formally incorporated within The University of Edinburgh.

References

1. Moray House Strategic Plan documentation, 1995.
2. Moray House Strategic Plan documentation, 1996.
3. Criteria for Mergers in Higher Education, Universities Funding Council, 1991.
4. *The Scotsman*, 22 June 1995, page 6.
5. University of Strathclyde, Post-merger evaluation, 28 September 1994, and University of Paisley, Post-merger evaluation, August 1995, Scottish Office Education Department.
6. Paper submitted to STEAC by T R Bone, 1985.
7. J P Arbuthnott and T R Bone, "Anatomy of a Merger", *Higher Education Quarterly*, Volume 4, 47, No 2, Spring, 1993, pp 103–119.
8. University of Glasgow/St Andrew's College, Merger Proposal, February 1998.
9. Report of Higher Education Quality Council, 1996.
10. Sir Stewart Sutherland, "Teacher Education and Training: A Study", Report number 10, *Higher Education in the Learning Society*, The National Committee of Inquiry into Higher Education, (Dearing/Garrick Report), HMSO, 1997.
11. Personal communication from Professor Tom Bone.
12. Statement by Principals of The University of Edinburgh and Moray House Institute of Education, 23 October 1996.
13. *The Times Educational Supplement Scotland*, Editorial, 25 October 1996.
14. *The Times Educational Supplement Scotland*, 11 July 1997.
15. Heriot-Watt University Press Release, 23 October 1996.
16. *The Times Educational Supplement Scotland*, 25 October 1996.
17. Personal letter to Professor Kirk from Professor MacFarlane dated 20 September 1996.
18. Statement by the Principal, Meeting of Senate, 30 October 1996.

19. *The Times Educational Supplement Scotland*, 25 October 1996.
20. Letter dated 20 October 1995 from Professor MacFarlane.
21. Letter dated 5 January 1996 from Professor MacFarlane to the Chairman of the Moray House Board of Governors.
22. Letter from Professor Kirk to Mr Wilson dated 13 October 1995.
23. Letter to the Principal of Heriot-Watt University from Moray House.
24. Letter to the Scottish Higher Education Funding Council and to the Scottish Office from Moray House.
25. Paper from the Principal to Moray House Board of Governors, 29 November 1996.
26. Report submitted to the March meeting of the Moray House Board of Governors (integration document).
27. Letter from the Principal of Heriot-Watt University to the Principal of The University of Edinburgh, 13 March 1997.
28. Letter from Professor Kirk to Professor Archer dated 18 November 1996.
29. *The Scotsman*, 19 March 1997.
30. Letter from the Chairman to Professor Archer dated 18 March 1997.
31. Press release by The University of Edinburgh, 19 March 1997.
32. Ibid.
33. A Proposal for Merger, February 1998.
34. A Proposal for Merger, February 1998.
35. Universities Funding Council, *Criteria for mergers in Higher Education*, 1991.

4

Retrospect and Prospect

It would be satisfying to claim that the transformation from independent monotechnic to Faculty of Education of the University of Edinburgh was the outcome of careful planning and the unfolding of a coherent strategy of institutional development. Superficially, from the standpoint of a detached commentator, it could be argued that each institutional move took Moray House a step nearer a predetermined target: from a CNAA-validated institution to an associated college, whose courses were validated by Heriot-Watt University; and from that "semi-detached" status to full incorporation within the University of Edinburgh. Each change of affiliation could be interpreted as marking the progressive enhancement of the institution's work to full university standing.

The evidence does not support that interpretation. Rather, two unrelated developments took place, each in response to changes in the context of higher education. Like other similar institutions, Moray House's efforts were directed towards accredited status with the CNAA, with the authority to award degrees while retaining its standing as an independent institution with its own board of governors. The threatened demise of the CNAA frustrated that aspiration and made it essential to establish alternative validating arrangements without any diminution in autonomy. The Heriot-Watt linkage secured that objective. Indeed, the ease with which agreement was reached with Heriot-Watt may be adduced as evidence of its appropriateness in the circumstances then prevalent. The merger with the University of Edinburgh, far from constituting a further step in an institutional plan, was profoundly disjunctive, involving, as it did, a switch of institutional allegiance, a very different mode of academic collaboration, and the ceding of institutional independence.

Moray House's institutional manoeuvrings were not, then, the product of rational planning, if that term implies the judicious scrutiny of options, the selection of a defensible line of development, and the careful shepherding of the institution along a predetermined path. On the contrary, Moray House's experience, like that of other institutions, involved exploiting, under pressure, opportunities that might protect its institutional mission and viability. The years between 1989 and 1996 involved a series of unsettling changes and institutions were encouraged to see themselves, not as participants in a collegial enterprise, but as competitors that were constantly on guard against predators, especially when these paraded as neighbours. Smaller institutions were particularly vulnerable and had to be capable of swift changes of direction if they were to thwart the ambitions of those committed to territorial aggrandisement. The overtures to Heriot-Watt University were a direct response to developments such as the appointment of the Principal of Napier to the Moray House Board of Governors, his apparent commitment to incorporating Moray House, with other Edinburgh colleges, as part of the Napier empire, and the apparent determination of the Scottish Office to restrict the institutional aspirations of the non-university institutions. All of these developments significantly heightened institutional insecurity. Six years later it was essential to respond to three pressures: SHEFC funding policy was a direct threat to institutional viability; the breakdown with Heriot-Watt made it essential to seek an alternative source of degrees for Moray House students; and the slowing down of Moray House's development suggested that it required the impetus of an intellectually invigorating kind. These three pressures combined to make merger an institutional imperative.

Of course, in entering into discussions with the University of Edinburgh, Moray House was motivated entirely by institutional self-interest. It was important to find a way of protecting, and if possible, perpetuating the institution's mission. There was no obligation to consider the impact of the Edinburgh discussions on Heriot-Watt University, which, after all, had been unsympathetic to a number of Moray House initiatives, most recently, its proposal to assume full responsibility for the promotion of senior staff. Indeed, Moray House's aggressive response to that initiative fully convinced staff and governors that Heriot-Watt did not share the types of excellence to which Moray House was committed.

If there was a degree of callousness in Moray House's decision to disengage from Heriot-Watt that is a reflection of the climate in which institutional self-interest had to be pursued with determination, if not ruthlessness. The extent to which institutional relationships had degenerated can be gauged by Heriot-Watt's reply to Moray House's decision to seek an alternative partner. It was understandable that there would have been disappointment at Riccarton and the need to mount a damage-limitation exercise was surely obvious. However, the repeated expressions of shock and dismay, and the pleas for participation in three-way discussions, had nothing whatever to do with the well-being of Moray House or its future: they had everything to do with protecting Heriot-Watt's self-interest and covering its understandable embarrassment. The stridency of Heriot-Watt's pursuit of its own interest was intensified the further the discussions between Moray House and the University of Edinburgh developed. The correspondence from the Riccarton campus, all of it concerned to marginalise the Principal of Moray House, was significant in this respect, for it apparently sought to destabilise the merger discussions and to justify Heriot-Watt's entitlement to a share of Moray House's work. In that correspondence reference was made to the possibility of "salvaging" something that would placate Heriot-Watt and to the desirability of achieving "a satisfactory split" of Moray House's work between the two universities. The callousness of such language was striking and served to reinforce the view that the interests of Moray House itself were of no concern to Heriot-Watt University. Appearing, as it did, on the very day on which the Moray House Board agreed to the Memorandum of Understanding with the University of Edinburgh and the broad principles governing the constitutional relationship between them, the Heriot-Watt press release ensured that the Board would reach a unanimous decision on both matters.

Fortunately, the University of Edinburgh remained impervious to the blandishments from Heriot-Watt University, and continued to confirm that the terms of its discussions with Moray House were on the understanding that the whole of Moray House, and not simply part of it, would merge with the University of Edinburgh. The University's constancy on this issue was a source of great encouragement at Moray House.

Of course, it would not inspire public confidence if it were to be thought that the provision of higher education in Scotland was the outcome of inter-institutional rivalry and personality conflicts. Fortunately, there are at least two levels of protection against the potential abuses of power when principals operate at or beyond the limits of their responsibilities. Whatever is negotiated in the clandestine demi-monde of institutional politicking has to be brought to public scrutiny through the various decision-making and consultation procedures that are deeply embedded in higher education institutions. Thus, both of the Moray House initiatives, with Heriot-Watt in 1989/90 and with the University of Edinburgh in 1996/97, were the focus of the most extensive consultation. On each occasion, the Board of Governors had before it at the key point of decision evidence of overwhelming support throughout the institution and, indeed, in the case of the Heriot-Watt linkage, support from many bodies and agencies in different parts of the country. A key element of that support on each occasion was provided by staff and student unions. Moreover, at the national level, inter-institutional alliances like mergers require the approval of the Secretary of State and of the Scottish Higher Education Funding Council. In reaching a decision, the Secretary of State is guided by a recommendation from the Funding Council, and in formulating advice the SHEFC invariably evaluates any proposal against a demanding set of criteria relating to the quality of educational provision and the most effective use of public funds.

The extent of the institutional consensus on merger with the University of Edinburgh could not disguise the fact that there were several areas of concern. As in other similar undertakings, the prospect of a change of employer created uncertainties, particularly, but not exclusively, amongst those on fixed-term contracts. The Transfer of Undertakings Protection of Employment (TUPE) regulations offered significant but not complete or comprehensive reassurance on employment matters and it was necessary, in consequence, to invoke goodwill and a sense of fairness in resolving difficulties that were undreamt of in

TUPE. These employment issues apart, there were reservations which related to institutional identity and mission. Many staff appreciated, despite the frequency with which the term was used, that Moray House had not embarked upon a *merger*. In the business world that term usually denotes the dissolution of two concerns and the establishment of a new legal entity. For all of the institutional alliances that have been forged in Scotland the term merger in that sense is inappropriate. They are more accurately seen as takeovers and, since they are invariably non-hostile, perhaps best described as takeovers by consent. That is, they are alliances that are asymmetrical: there is always a senior partner which, whether or not it initiated discussions, is seen as incorporating a smaller partner. In such circumstances, it is understandable that staff of the smaller institution should be concerned about the preservation, within the enlarged institution, of those values which command their allegiance and govern their approach to higher education and the conduct of academic life.

At Moray House, some colleagues expected that the merger would entail that teaching and professional activities would become submerged within the culture of a research-led university. In the process of planning the new institutional framework it was essential to acknowledge that, whether or not these fears were justified, they were certainly experienced, and it was important to ensure that they would be allayed. The terms of the merger agreement sought to provide reassurances on all of these matters.

In the first place, it was argued that the designation *Moray House Institute of Education* would be retained, not as an assertion of independence but to symbolise the continuation of an approach to professional education that was internationally renowned. Constitutionally, there would be a faculty of education in which Moray House's activities would be incorporated, and the faculty would operate in full accordance with the codes of academic and administrative practice of the University as a whole. That is, the retention of the title in no way threatened the integrationist principles on which the whole merger was based.

Secondly, while the merger involved the full incorporation of Moray House within the University and therefore the adoption of University policies and procedures, it was acknowledged that, in cases where institutional practices differed, the principle of best practice would be adopted. In line with that agreement, the University saw much to commend in the Moray House Modular Master's Scheme and has incorporated the credit accumulation and transfer model, with all the flexibility it confers, into its postgraduate provision. In addition, in several contexts since the merger former Moray House practices have been adopted, or recommended for adoption, by the University community as a whole.

Thirdly, while the merger document acknowledged the need to extend research activity in the Faculty of Education, it recognised that, given the heavy demands of programmes for teaching, community education and other fields, and given also the need to ensure that members of the Faculty of Education continued to support practising professionals, it would be unreasonable to expect the Faculty of Education to match the volume of research participation evident in the established faculties in the University. In the years prior to merger Moray House had been committed to the growth of a research culture, believing that, as a major centre of professional education, it had an obligation to contribute to the advancement of knowledge as well as its dissemination, and believing also that strong courses flourish in a culture of critical enquiry and debate. The merger was seen as a way of accelerating the growth of research, not least by incorporating the University's established researchers and education specialists within the new Faculty. The Faculty's research strategy, reinforced by significant investment in new appointments by the University, will ensure that the research profile of the Faculty is enhanced with a progressive increase in the proportion of active researchers.

Fourthly, the University has affirmed, if that were necessary, the importance of teaching to its life and its commitment to the enhancement of the educational experience offered to students. Whatever assumptions were made by universities about the importance of teaching, there is no doubt that the advent of quality assessment and the work of the Quality Assurance Agency for Higher Education have made it an institutional imperative for universities to give priority to the quality of the learning experience they provide and to the enhancement of the academic standards to which their awards attest. The University of Edinburgh is similar to other institutions in this respect. Indeed, one of the attractions of the merger from the University's point of view was that it would incorporate a large number of staff who had pronounced expertise in teaching and for whom teaching was an important source of professional satisfaction.

Finally, if there were fears about the extent to which the Moray House commitment to professional action would be maintained, these have been dissipated. In all of the Institute's undergraduate programmes, and in line with the guidelines issued by the Secretary of State and the expectations of professional bodies, placement activities have a centrally important place. The merger document strongly affirmed this commitment. It stated:

> *It is essential that programmes of education are as academically demanding as others offered by the University . . . At the same time, it has to be recognised that the professions for which the existing Moray House trains expect that graduates will be accomplished practitioners. While it is crucially important to ensure that professional skills are firmly grounded in theoretical understandings, it is also vital that theoretical understandings can issue in competent professional performance. Moreover, in professional programmes, both theoretical understanding and professional performance should be reflected in the final award that is conferred.*

On this important matter the University recognised the principle of complementarity: that there is, indeed, a higher education perspective to be brought to professional education, which is the product of reading, research and reflection, and which has to be seen in relation to the insights that come from the experience of the seasoned practitioner. The University recognises that staff in an institution of higher education ought to be able to draw on the first of these sets of expertise and that it is the function of schools and other agencies to provide the second. The merger document articulated that stance in these words:

> *It is now common ground that schools and other agencies on the one hand and higher education institutions on the other have distinctive and complementary contributions to make to the education of teachers and related professionals. . . . It is through engaging in work with learners that students' growing professional understanding is developed. It is in the classroom and the community agency that students are able to access the "craft" knowledge of the accomplished professional, the knowledge that is grounded in cumulative experience. However, the professional educator does not simply practise a craft: the work is underpinned by an extensive range of theoretical understanding. It is the function of the university to engage students in the analysis of the theoretical underpinnings of teaching, to familiarise them with the insights derived from research and from the academic literature on the theory and practice of education. The challenge is to find a way in which these two distinctive but complementary knowledge bases are to be co-ordinated to form a coherent programme of professional preparation. The kind of collaborative partnership envisaged is one in which the university works in close association with a network of federated schools and community agencies, sustaining, through joint staff initiatives, co-operative teaching, and academic interchange of various kinds, a culture of professional co-operation and development, recently echoed in the Sutherland Report. Through these various points of contact and joint activity, the enlarged University of Edinburgh can become a partner in the revitalisation of the work of schools and other agencies, and practising professionals come to enjoy the benefits of engaging in the life and work of the University.*

The success of the merger in sustaining and enriching the Moray House tradition of professional education will be judged by the extent to which it is able to realise that aspiration.

The evidence on the effectiveness of the merger to date is encouraging and there are four sources of evidence which may be invoked to justify that assessment. First, the various objectives that had been set by the merger were all overtaken within the deadline. Indeed, the relocation of Cramond-based activities to the city centre took place one year ahead of schedule. Secondly, Dr Chris Masters, Chair of the Funding Council, in his letter to Wendy Alexander, the Minister responsible for higher education, in June 2001, stated

> *From the evidence of the self-assessment document and the meetings that the Mergers Committee has held at the University, the Council considers that this has been a successful merger.*

Thirdly, the Research Assessment Exercise (RAE) of 2001 resulted in a rating of 4 in Education, and placed Edinburgh's Faculty of Education as the fourth largest concentration of educational researchers in the UK. The rating of 3A in Physical Education was maintained, despite an ambitious programme of diversification to ensure that the Scottish Centre for Physical Education, Sport and Leisure Studies develops strength across the full range of specialism in these fields. The infrastructure established and the investment made by the University in this area of work seek to secure a centre of international standing.

Fourthly, in the course of session 2001/2002, the Faculty participated in the piloting of a new approach to assuring the quality of provision in initial teacher education. The report from the *Collaborative Review* evaluation team – drawn from HM Inspectorate, the General Teaching Council, and the Quality Assurance Agency – demonstrated *Confidence* in the standard of provision and the achievements of students. Besides, the Faculty's teaching provision was judged *Commendable* in all four domains. Moreover, five aspects of the Faculty's provision were judged to be *Exemplary*, in the sense of being worthy of emulation in other institutions.

In the course of 2001/2002, the University of Edinburgh undertook a major restructuring exercise of its academic organisation. From August 2002 Faculties will disappear and the academic activities of the University will be located in twenty Schools, grouped into three major Colleges. The Faculty of Education will become The Moray House School of Education. A secure foundation for the development of that School has been laid. Long may it flourish.